Vehicle Maintenance and Repair Series

Vehicle Maintenance

Vehicle Fitting Units Level 1 and 2

Roy Brooks, Jack Hirst, John Whipp

THOMSON ™

Australia • Canada • Mexico • Singapore • Spain • United Kingdom • United States

THOMSON

Vehicle Maintenance: Vehicle Fitting Units
Level 1 and 2, 2nd edition

For more information, contact Thomson Learning, High Holborn House, 50-51 Bedford Row, London, WC1R 4LR or visit us on the World Wide Web at: http://www.thomsonlearning.co.uk

British Library Cataloguing-in-Publication Data
A catalogue record for this book is available from the British Library

ISBN-13: 978-1-86152-808-7
ISBN-10: 1-86152-808-6

First edition 1997 Macmillan Press Ltd
(Vehicle Maintenance Service Replacement Level 1)
Second edition 2001 Thomson Learning
Reprinted 2004 and 2006 by Thomson Learning

Typeset by Wearset, Boldon, Tyne and Wear
Printed in Croatia by Zrinski d.d.

Acknowledgements

The editor, authors and publishers would like to thank all who helped so generously with information, assistance, illustrations and inspiration. In particular David Bull, Harvey Dearden, Steve Green, David Thistlethwaite and staff of the Motor Vehicle Section of the Burnley College, together with the firms and organisations listed below. Should there be any errors or omissions, they are completely unintentional.

A C Delco, Alfa Romeo (Great Britain), Associated Tyre Specialists Ltd, Autogem Ltd, Bosal UK Ltd, Robert Bosch Ltd, Butterfield Equipment Ltd, Citroen UK Ltd, Department of Transport, Alexander Duckham & Co Ltd, Firemaster Extinguisher Ltd., Focal Displays Ltd, Ford Motor Co Ltd, General Motors, Goodyear Great Britain Ltd, Hofman Balancing Techniques Ltd, Hofman Werkstatt-Technik Gmbh, Koni, CCS Component Distribution, Kwik-Fit (G.B.) Ltd, Lucas Industries PLC, Lythgoe Motor Group, Mercedes-Benz UK Ltd, Michelin Tyre Plc, Mitsubishi Motors (The Colt Car Co Ltd), Moldex, Motorway Tyres and Accesories Ltd, National Tyre Services Ltd, Nissan Motor (GB) Ltd, North Safety Products, Peugeot Motor Co Plc, Pirelli Tyres Ltd, R.A.C., Renault UK Ltd, Renault UK Ltd (Manchester), Saab Great Britain Ltd, Sachs Automotive Components Ltd, Safety Equipment Centres, Sanderson Ford Ltd, Seton Ltd, Siebe Garage Equipment Group, Signs and Labels Ltd, The Ian Skelly Group, S P Tyres UK Ltd (Dunlop), Sykes-Pickavant Ltd, Radio Shack Division of Tandy Corporation, Rover Group Ltd, Tecalemit Garage Equipment Co Ltd, Alfred Teves Gmbh, Thistlethwaites Tyres and Batteries, Tyre Industry Council, Tyvek Protective Clothing, Vale Garage Ltd, Vauxhall Motors Ltd, Verdestein (UK) Ltd, Volkswagen Group United Kingdom Ltd, J W Whittle Ltd.

Every effort has been made to trace copyright holders, but if any have been overlooked the publisher will be pleased to make the necessary arrangements at the first opportunity.

Contents

This book contains basic **Essential Knowledge** required for the following MOTOR INDUSTRY TRAINING COUNCIL (MITC) Qualifications in:

VEHICLE FITTING Level 1 and Level 2

Fast Fit Technician
Specialist Tyre Fitter
Auto Electrician

VEHICLE MAINTENANCE and REPAIR Level 2

Vehicle Inspector
Light Vehicle Technician
Heavy Vehicle Technician

The Book also provides a very suitable introduction for any person who is seeking employment in, or who is about to enter, the Motor Trade.

Preface

Welcome to the retail motor trade

Who knows, you – yes you, the person who is reading this book – could well be a future racing or rally car technician, or even a budding driver. They all had to start somewhere – draining oil, changing tyres, removing exhausts, or even sweeping up the workshop!

Vehicle Maintenance is your introduction to the varied and exciting world of automobiles. It aims to steer you on the pathway to success.

This book guides you through many of the practical jobs you will have to do, but vitally helps you to look after your own health and safety. Where necessary, it also gives the basic theory – so you can understand the reasons why.

Enjoy yourself with *Vehicle Maintenance* – it is not just for reading. In its pages you will find activities and questions that are stimulating and rewarding. They will help you to observe and remember important points. Make the book a valuable source of reference by filling in the answers to the questions.

The important qualifications of NVQ/SVQ Vehicle Fitting, Levels 1 & 2, and the needs of other similar courses, are fully covered. This book will help to provide evidence for your portfolio. If you want to go on to gain further qualifications, the more advanced books in this series are waiting to help you. Good luck!

Roy Brooks
Series Editor

Coverage of standards, by unit

QUICK CHECK
UNIT GRID

VEHICLE
MAINTENANCE

VEHICLE FITTING
LEVELS 1 & 2
VEHICLE MAINTENANCE
and REPAIR
LEVEL 2

Subject material in chapters covers **Basic Essential Knowledge** for the unit areas indicated.

UNIT NUMBERS and TITLES	**CHAPTERS**	1. Safety and good housekeeping	2. Working relationships	3. Communication	4. Handling and merchandising stock	5. Tyres	6. Electrical	7. Brakes, suspension, steering	8. Exhaust, cooling, clutch, lubrication
1 Contribute to Good Housekeeping		●	●	●	●	●	●	●	●
2 Ensure Your Own Actions Reduce Risks to Health and Safety		●	●	●	●	●	●	●	●
3 Maintain Positive Working Relationships		●	●	●	●	●	●	●	●
4 Identify Faulty Components		●	●	●		●	●	●	●
5 Remove and Replace Components		●	●	●	●	●	●	●	●
6 Carry Out Tyre Related Adjustments		●	●	●		●			
7 Repair Tyres		●	●	●	●	●			
8 Locate Simple Electrical/Electronic Faults		●	●	●			●		
9 Remove and Replace Electrical Components		●	●	●	●		●		
12 Diagnose Non-Complex System Faults		●	●	●		●	●	●	●
18 Identify and Agree Customer Vehicle Needs		●	●	●		●	●	●	●
19 Inspect Vehicles		●	●	●		●	●	●	●
20 Valet Vehicles		●	●	●	(See Maintenance and Repair of Road Vehicles L2 for Valeting text)				
I Contribute to the Security of the Workplace (DNTO)		●	●	●	●	●	●	●	●
I Solve Problems for Customers (Customer Service)		●	●	●	●	●	●	●	●
I Place Goods and Materials in Storage (DNTO)		●	●	●	●				

I = Imported Unit

Chapter 1

Safety and good housekeeping

In this chapter you will learn about:

- **health and safety at work** – what to be aware of
- **accidents and first-aid** – what to do if something happens
- **hazards** – what to watch for
- **personal protection** – clothes and equipment
- **safe handling** – of loads, of equipment, and of harmful substances
- **health and safety signs** – what they mean
- **emergency procedures** – fire alarms and fire extinguishers
- **good housekeeping** – keeping the workshop clean, tidy, and safe
- **workshop resources** – looking after equipment, power, and time

It is vital for your own health and safety that you keep the place where you work clean and tidy. Doing this will be part of your job.

This chapter explains how to avoid accidents and how to keep healthy at work. It tells you what to do in an emergency. It also explains how to get rid of waste, some of which may be dangerous to others.

The chapter looks at security, at how to move loads without hurting yourself, and at how to use resources economically.

Health and safety at work

You must **always take care of your own health, hygiene and safety**. This chapter looks at health and safety matters that affect you at work. It also explains some of the regulations. These exist to make sure that you work in good, safe conditions. All recent major regulations have developed from the legislation in the **Health and Safety at Work Act 1974**.

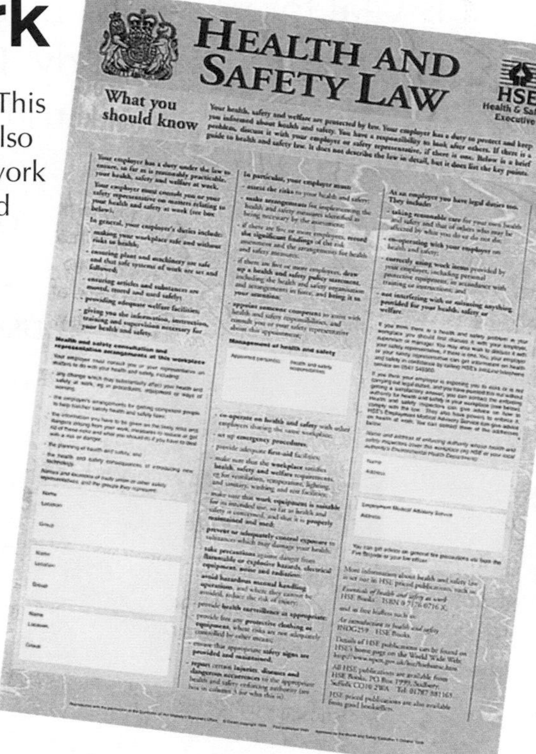

Regulations

In the United Kingdom and Europe, health and safety matters are covered by the **Management of Health and Safety at Work Regulations 1992**.

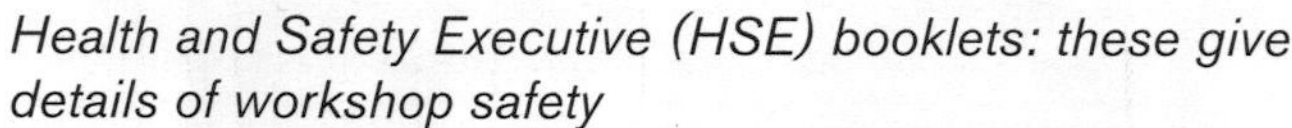

Health and Safety Executive (HSE) booklets: these give details of workshop safety

As you train you will need to learn and understand the safety regulations that apply to your job. **You are as responsible as your employer for following workshop safety regulations**. The Health and Safety Law Poster **'What you should know'** is shown on this page. It states employer and employee duties within the modern framework of health and safety law. You will find copies of this poster and others displayed in the workshop. For your own safety, read them.

Duties

Under the Health and Safety at Work Regulations, *you* have certain duties. Bear these in mind as you work.

- Take reasonable care of your own health and safety.
- Take reasonable care for the safety of other people who may be affected by your actions.
- Work with your employer to keep safety rules.
- Report any accidents, hazards, or damage to equipment.

HEALTH AND SAFETY AT WORK ACT 1994

Trainees and Students have the following responsibilities under the Health and Safety at Work Act.

1) To have personal responsibility for the health and safety of self and others with whom they are working.

2) To observe safe standards of behaviour, dress and protective clothing as required by the Centre/College Policy.

3) To use, and not to wilfully misuse, neglect or damage, nor interfere with devices and equipment etc. provided for his or her health and safety.

4) To assist by reporting to the instructor or lecturer, any hazards, defects, inadequacies or dangers which he or she considers exists in his or her workplace.

5) To ensure that any accidents or dangerous occurrences are reported as soon as possible.

A Smith
Principal

A typical training centre or college health and safety notice

Warning

If you do not follow the Health and Safety at Work Regulations, you could be taken to court. For example, suppose you were welding an exhaust pipe. If you did not wear the goggles provided, you would be breaking the law, because you would not be taking reasonable care of your own health and safety.

Accidents

An **accident** involves something that is unexpected and unplanned. One or more people may be injured.

If the accident is minor, it may just be inconvenient. If it is serious, though, it could affect you for the rest of your life.

Accidents may be caused if you:

- do not know the dangers involved in what you are doing
- daydream
- do not take enough precautions
- fool around.

Accidents may also occur because of faulty equipment or bad work conditions:

- unsafe tools
- unguarded machinery
- poor ventilation
- poor lighting.

ACTIVITY

List three other causes of workshop accidents.

1 ....

2 ....

3 ....

HEALTH & SAFETY

Remember: **all the time, be alert to prevent accidents**.

In any job, be aware of:

- the dangers
- available protection.

First-aid

In your place of work, someone will have been specially trained as a first-aider. Normally this person will give first-aid. If the accident is minor, you may be able to help.

Whenever an accident happens, it is important to record the details in a special **accident book**.

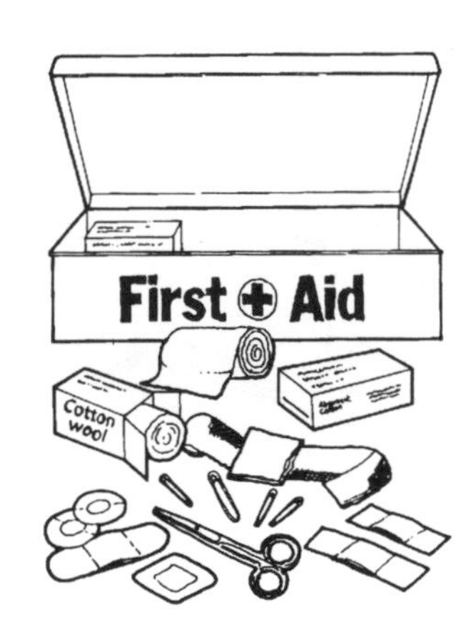

TIP

Every first-aid kit should contain a card giving advice on what to do if anyone is injured.

Do not wait for an accident, though. Get training in first-aid as soon as you can.

ACTIVITY

Can you answer these questions? If not, find out.

1 In an accident involving a person and damage to a vehicle and garage equipment, what should you attend to first?

....

2 Who in your workshop is a competent first-aider?

....

3 Where is the first-aid kit kept?

....

4 Where is the accident book kept?

....

Hazards

A **hazard** is anything that might cause an accident or injury.

Look around your workshop. You will probably see at least one hazard, perhaps more. Some can quickly be removed. Examples are:

- wheels left lying on the floor, after being taken off a vehicle
- oil spilt on the floor
- a trolley jack handle left lying where someone could trip over it.

Some hazards are always present. Warning notices or guards, or both, keep the risk to a minimum. Examples are:

- brake tester rollers
- a wheel balancer
- a stand drill.

HEALTH & SAFETY

Hazard
'a hazard is something with potential to cause harm'.

Risk
'a risk is the likelihood of the hazard's potential being realised'.

ACTIVITY

Look at the garage workshop layout shown. Draw a ring round each hazard that you see. (You should ring about 12.)

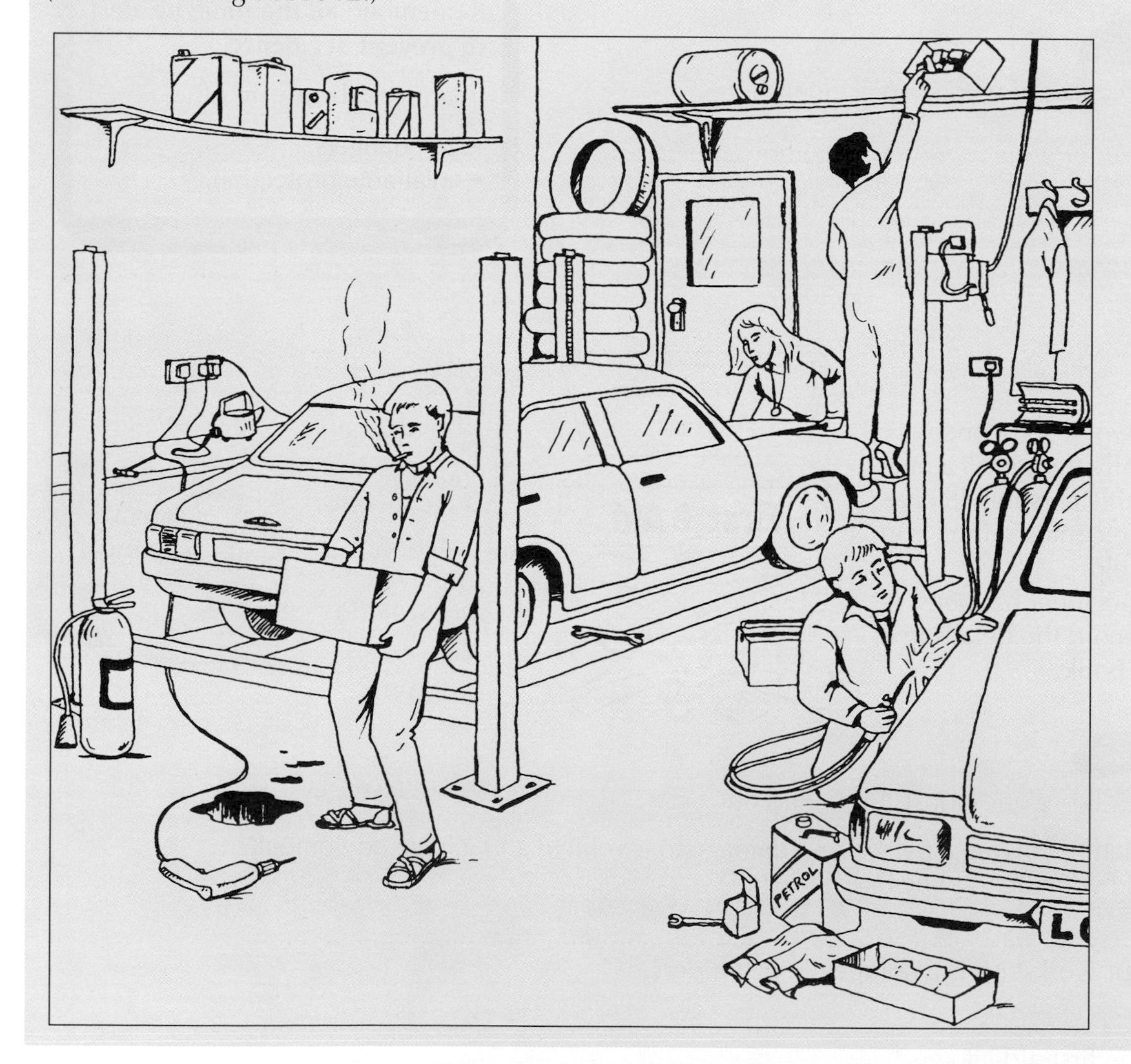

WORKSHOP

1 Next time you are in the workshop, look round to see what hazards are present.
2 How could they be avoided?

Personal protection

Various equipment is available which can be worn or held by people at work, protecting them from risks to health and safety. This equipment is covered by the **Personal Protection Equipment at Work Regulations 1992**. All Personal Protective Equipment (PPE) in use at work should carry the CE mark and where appropriate should comply with a European Norm (EN) standard.

HEALTH & SAFETY

Your personal presentation at work should:

- help to ensure the health and safety of yourself and others
- meet legal requirements
- be in accordance with workplace policies.

Personal presentation

Take care in what you wear.

To protect yourself and your clothes, it is sensible to wear:

- one-piece overalls (a boiler suit)
- stout footwear (preferably with steel toe-caps)
- a suitable cap or bump cap.

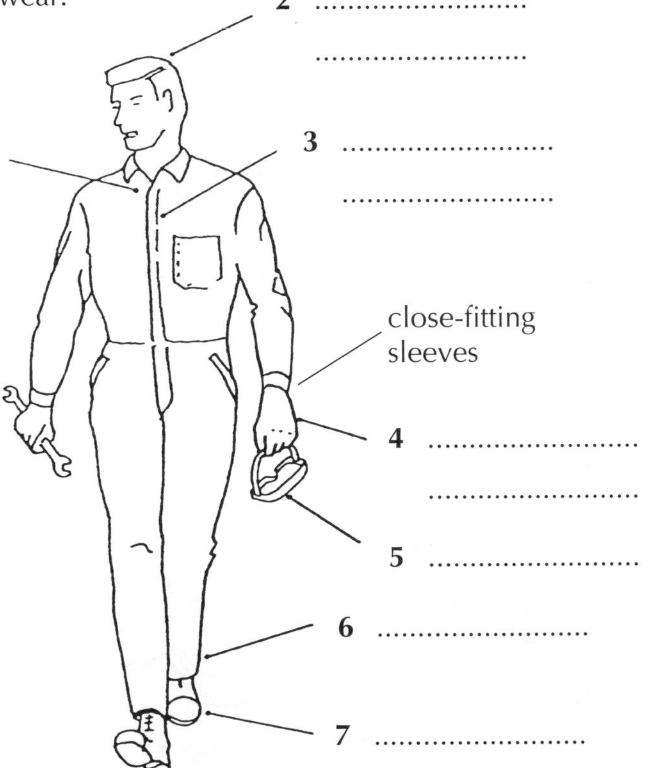

Do **not** wear:

- loose or torn overalls (especially if the *sleeves* are loose or torn)
- rings or watches
- trainers or similar
- long hair (unless protected by suitable headgear).

Specialised personal protection equipment

In your work you will need specialised equipment to protect you. Examples include dark-tinted glass goggles when welding, masks when painting, waterproof clothing when steam cleaning, or high-visibility clothing when going out to vehicle breakdowns.

As you do vehicle repair and maintenance work you will sometimes need special equipment to protect the top of your head, your eyes, your ears, your hands and feet, and your breathing.

ACTIVITY

Check this technician for safety features. Write down the names of these features.

HEALTH & SAFETY

Large rings, a fancy watch, and gold chain might be good to wear for a night out. If they get tangled with machinery though, at work, you could end up in hospital.

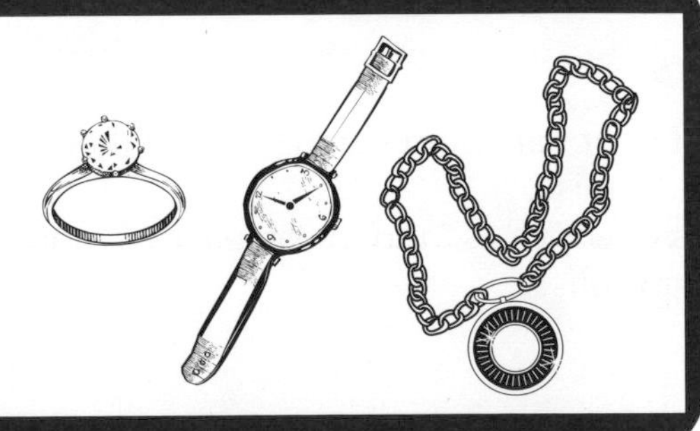

ACTIVITY

Look at this list of protection equipment. What hazards would each item protect you from?

Protection	Type of hazard
Bump hat	1
Dark goggles	2
Face mask	3
Earplugs	4
Gloves	5
Wellingtons	6

Safety caps

Bump caps protect your head from banging on the underside of a vehicle when you work under a ramp.

Soft caps keep your head and hair clean. They also prevent long hair from catching in revolving parts, such as drills on the bench, or engine drive belts under the bonnet.

Bump cap

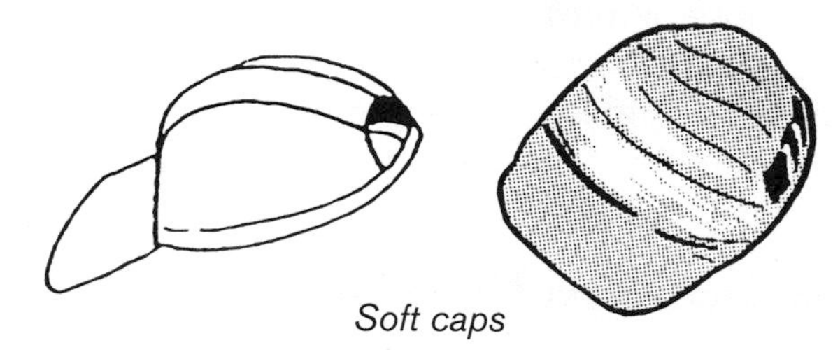

Soft caps

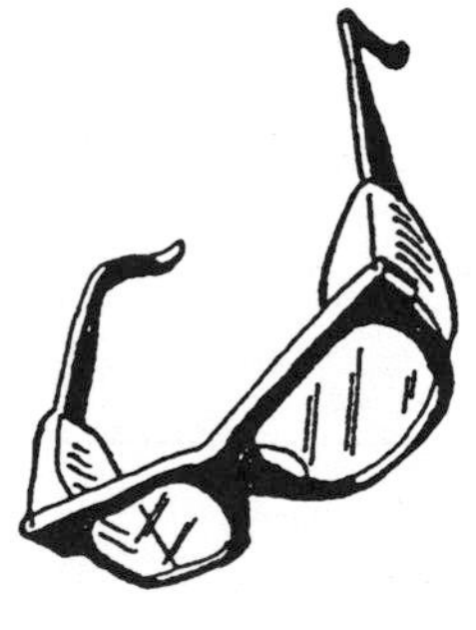

Spectacles

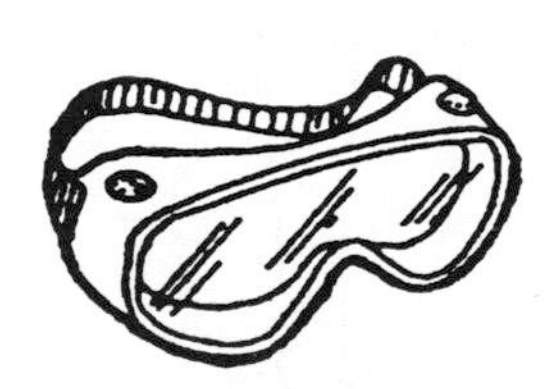

Goggles

Welding goggles

Eye protectors

An accident to the **eyes** can be very painful and may result in blindness.

Spectacles protect you from rust or dirt falling off the car. Some have side shields and can be adjusted to fit your face.

Goggles protect you from dust and chemicals. They are used, for example, when sanding body filler off bumped car wings.

Welding goggles protect your eyes from the bright glare of the welding flame.

Ear protectors

Ear muffs protect your ears from damage when there is a loud, continuous noise.

Earplugs are as effective as, and in some cases more effective than, ear muffs!

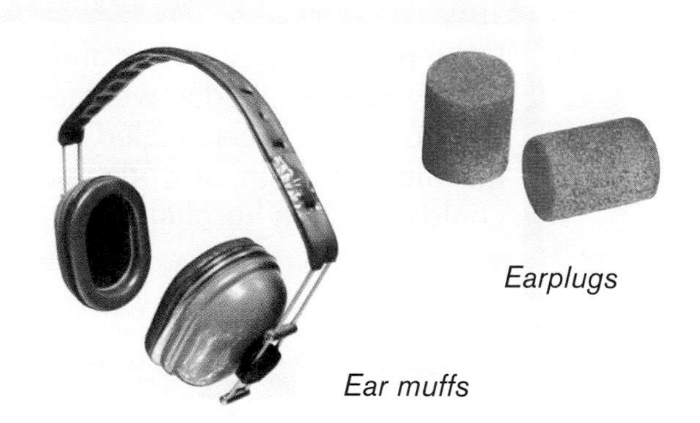

Earplugs

Ear muffs

Masks

Face masks protect your lungs from dust. (Some of the dust may be toxic.) They use special moulded pads, made from cotton gauze or special filter paper.

Gas respirators are used in vehicle paint shops. The paint fumes may be toxic.

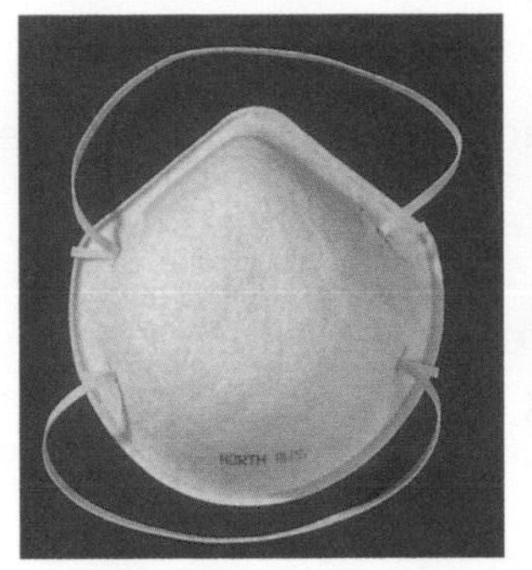

High grade dust mask

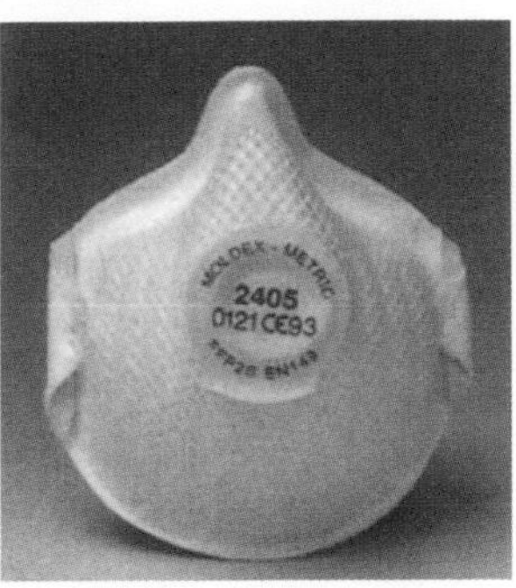

Valved, dust/mist mask

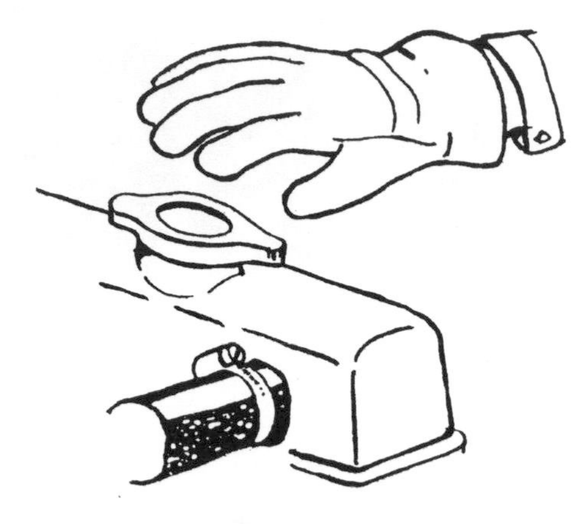

Heat-resistant gloves

Hand protection

Industrial gloves should be used when moving rough or heavy parts. They protect your hands and wrists from cuts, scratches and burns.

Heat-resistant gloves should be worn when working on items such as a hot exhaust or radiator.

HEALTH & SAFETY

Take care – gloves, or your hands, can easily get caught in revolving parts.

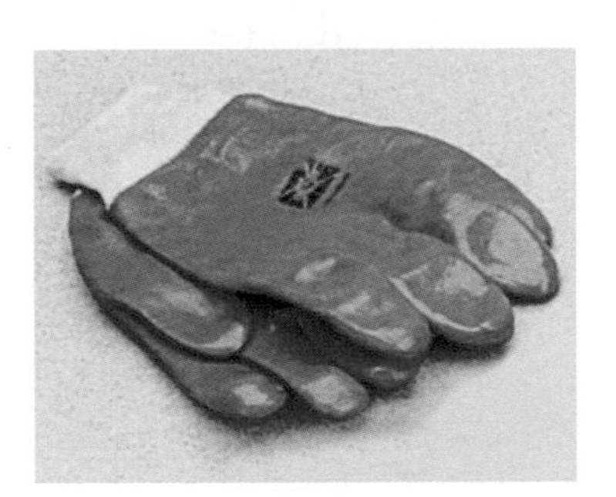

Plastic-coated gloves

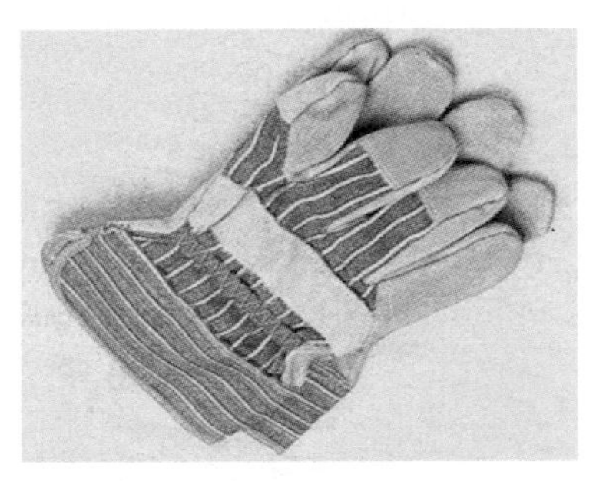

Leather-faced fabric gloves

Skin protection

Before you start work, put **barrier cream** on your hands. If you have sensitive skin you might easily develop an infection such as **dermatitis**. Thin plastic gloves (like surgical gloves) can be worn to prevent contact with fuel and oil.

After work, clean your hands with an **antiseptic hand cleaner**. Rub this on your hands *before* you get them wet.

Keep your overalls clean, and do not put dirty rags in your pockets. Oil might pass through your clothes and onto your body, and you might develop a skin infection.

HEALTH & SAFETY

If your hands become sore, cracked, itchy and red, see a doctor. You might have dermatitis.

Foot protection

Safety boots protect your feet and toes from falling objects. In a workshop there is also a risk that a car might run over your feet!

Steel toe-cap boots

Total protection

Total waterproof protection will sometimes be needed.

- When working at a car wash or valeting firm, wear waterproof clothing.
- When working on breakdowns, wear high-visibility clothing.
- When paint-spraying a car, wear Tyvek overalls. These have elasticated hoods, cuffs and ankles. Wear a gas respirator, too.

Safe handling

Moving loads

A **load** is any heavy object that must be moved, whether by hand or by lifting equipment. Heavy loads may be moved in the ways shown.

QUESTIONS

1 Name each type of transport.

a .... **b** .... **c** ....

2 Only one unit is loaded correctly. Which is it? ....

Correct handling techniques

When you have to lift something big or heavy, you need to lift it in the right way. Look at the drawings and read the numbered instructions.

Lifting a heavy part from the floor:

1. Stand as close to the load as possible. Spread your feet.
2. Bend your knees and keep your back in a straight line. Do not bend your knees fully, as this would leave you with little lifting power.
3. Grip the load firmly.
4. Raise your head.
5. Lift by straightening your legs. Keep the action smooth.
6. Hold the load close to the centre of your body.

Unloading onto a bench:

1 Bend your knees to lower the load. Keep your back straight, and the weight close to your body.
2 Be careful with your fingers as you set the load down.
3 *Slide* the load into position on the bench. Push with your body.
4 Make sure that the load is secure, and that it won't tip, fall or roll over.

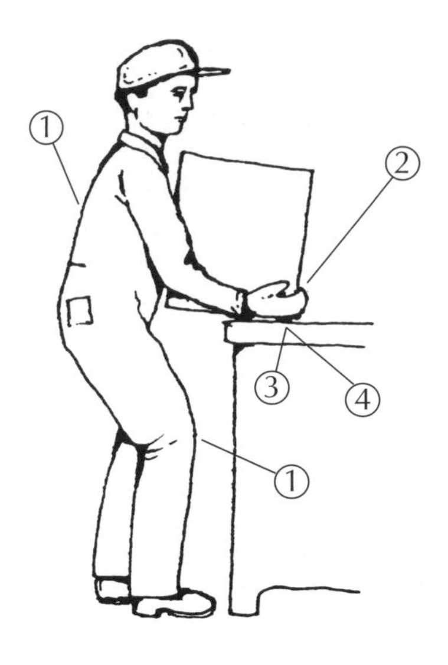

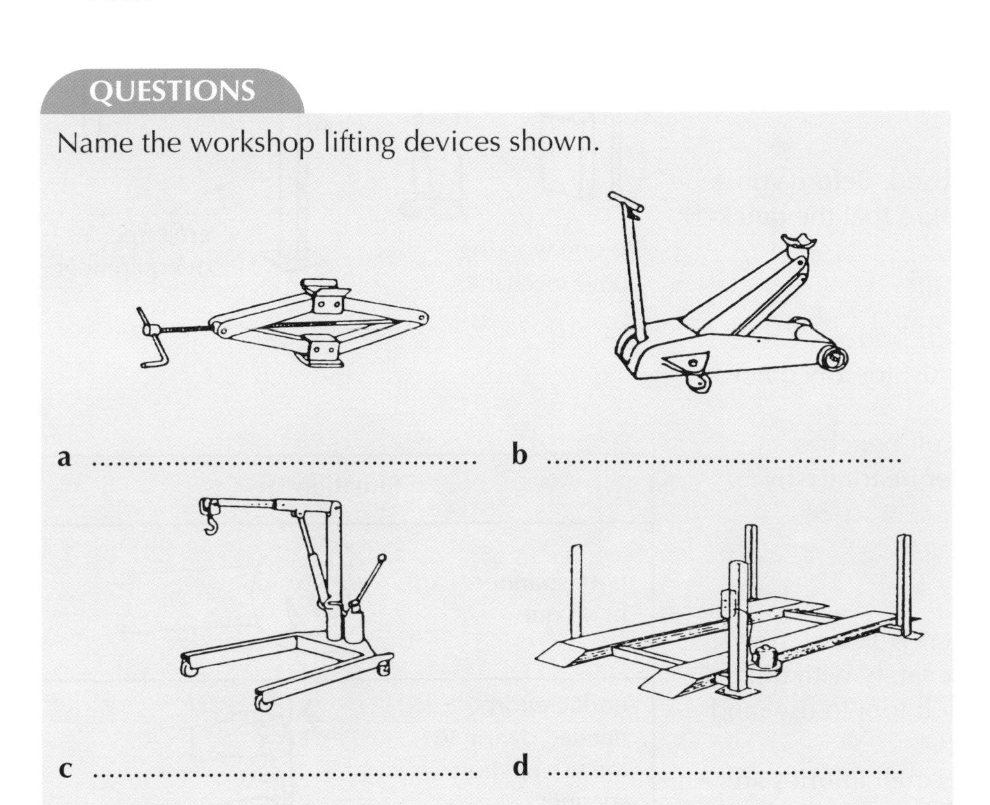

QUESTIONS

Name the workshop lifting devices shown.

a .... b ....

c .... d ....

Safe use of garage machinery and equipment

Ramps and jacks

In the workshop you will often use **vehicle lifts** and **trolley jacks**. Make sure you know how to use them safely.

QUESTIONS

What precautions should you take before working on a car in this condition?

1 ....

2 ....

3 ....

4 ....

The diagram opposite shows basic precautions you should take when working under a lift. Here are three extra precautions:

- Do not exceed the lift's **safe working load** (**SWL**).
- Before raising a car, check that the radio aerial, bonnet and boot lid are down. They could hit lights, beams or the roof.
- Before lowering, make sure that all tools and old parts have been removed.

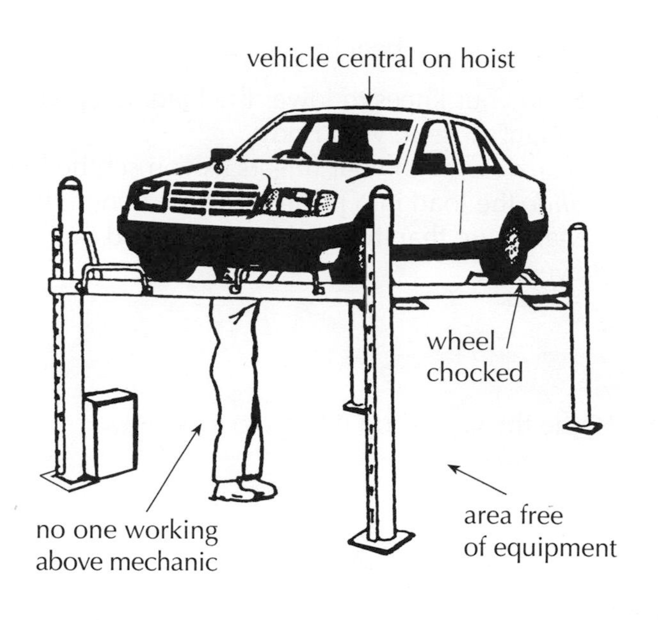

Compressed-air equipment

Compressed air is dangerous if misused. Before you use flexible pipe extensions, make sure that the **quick-release couplings** are fully engaged.

When working with compressed air:

- never direct it onto any part of your body
- never use it to blow away brake dust (or any other type of dust)
- never use it to clear dirt or filings off benches
- never use it to clean ball and roller bearings (by spinning them).

Hand tools

Hand tools are spanners, sockets, screwdrivers, pliers, hammers, chisels and files. To work safely with them use your common sense, know which tool to use, and follow safe procedures.

In a workshop the most common small injuries are cut fingers or skinned knuckles and fingers. Usually these are due to the misuse of a hand tool. The diagrams opposite show situations where this can occur.

Situation
poor spanner fit on nut
undue effort needed owing to the use of short spanner
using a file without a handle
using a blunt screwdriver

Rotating machinery

All high-speed rotating machinery, such as wheel balancers, drills and grindstones, should have **guards** fitted.

HEALTH & SAFETY

Motor vehicle repair work is dirty. **Cuts, however minor, should always be cleaned and treated straight away.**

Handling harmful substances

Workshops often store dangerous **chemicals**. Some could catch fire; some could even explode when handled. Others are **corrosive** or **caustic**, and could damage your skin.

There are regulations about the **Control of Substances Hazardous to Health** (**COSHH**). These state that every hazardous substance must be described on a health and safety data sheet. The sheet gives details of safe handling, and says whether protective equipment should be worn.

HEALTH & SAFETY

If a chemical is hazardous, this is usually stated on the packaging. Here are some of the standard warning symbols.

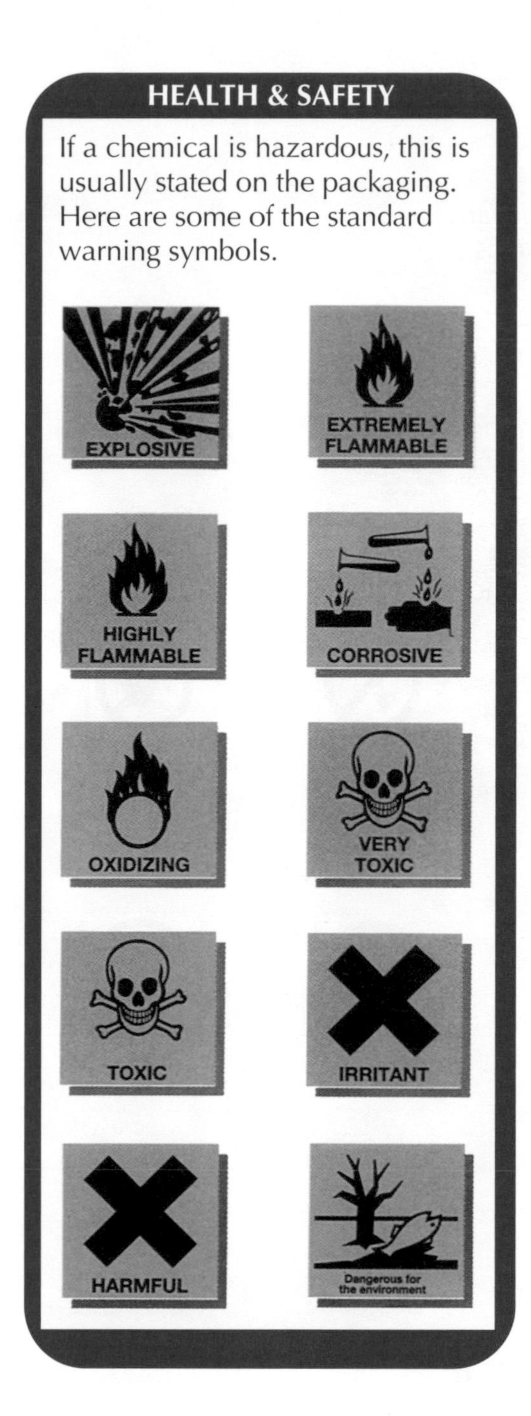

DATA SHEET No. 14 DECEMBER 1989

PART No. BC1 BRAKE CLEANER 500ml AEROSOL - PACK 12
BC2 BRAKE CLEANER 500ml AEROSOL - PACK 1
BC3 BRAKE CLEANER 5L C/W TRIGGER SPRAY - PACK 1

PHYSICAL FORM
Clear colourless liquid.

CHEMICAL COMPOSITION
Trichloroethane 111. a non-flammable chlorinated hydrocarbon solvent.

STORAGE
Pressurised container - store away from heat and direct sunlight.
Bulk - store in closed containers as supplied.

SPILLAGE
Small spillages will soon evaporate in well ventilated areas.
Large spillages can be recovered for reuse by distillation.

PERSONAL PROTECTIONS
Avoid excessive skin contact.

ADDITIONAL PRECAUTIONS
Ensure adequate ventilation whilst in use. Harmful by inhalation and if swallowed. In common with other solvents of this type try to ensure use area is well ventilated, as build-up of vapour may produce drowsiness.

HAZARD INFORMATION - COSHH REGULATIONS
Contains 111 Trichoroethane <70% TLV 350 ppm. LD50 (rat) 10 - 12 Kg
Contains butane <30% TLV 600 ppm. (Aerosol only).

SAFETY FIRST
NOT TO BE TAKEN INTERNALLY
KEEP OUT OF REACH OF CHILDREN
DO NOT PUNCTURE OR INCINERATE AEROSOL EVEN WHEN EMPTY
KEEP AWAY FROM DIRECT HEAT

FIRST AID
EYES- Rinse splashes to eyes with water for several minutes. Seek medical aid if irritation persists.
SKIN- Wash with soap and water. Apply soothing cream if necessary.
INGESTION- Wash out mouth with water. Seek medical aid immediately.
INHALATION- Remove patient to fresh air and seek medical aid.

An example of an AUTOGEM health and safety data sheet

ACTIVITY

Find out which dangerous substances are kept in *your* workshop.

Combustible materials

Some liquids or chemicals found in a garage catch fire very easily. Petrol is the most obvious example. Vapour from such chemicals could be ignited by a spark – even the tiny spark when a light switch is operated.

Such fluids must be stored in **fireproof containers**. These are designed to prevent leakage and evaporation.

HEALTH & SAFETY

For your own safety, make sure the equipment you use has been checked and is safe.

Electrical safety

The main dangers caused by **electricity** are:

- **fire** – due to cables being overloaded, overheating, or loose connections
- **electric shock** – due to touching a live circuit.

Unguarded cables or connections, like those in the diagram opposite, could cause a fire or a shock.

In industrial premises, all electrical equipment must be checked regularly by a qualified person.

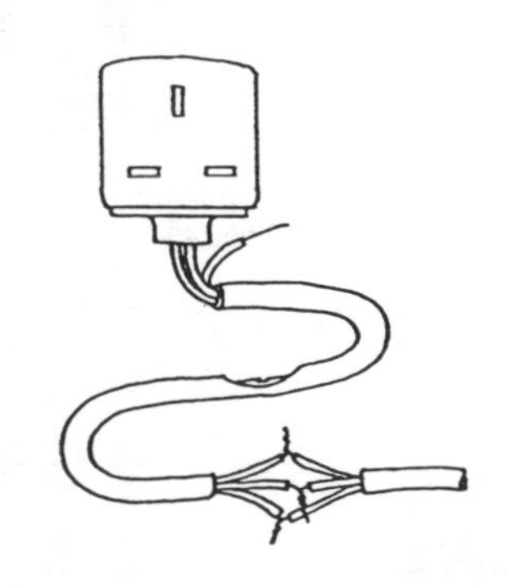

Health and safety signs

All public places, including workshops, must display safety signs to warn people of dangers. If you look around buildings you will see such signs. Some will be so familiar that you hardly notice them.

By law (**BS 5378** and **BS 5449**) safety signs must clearly show what they mean. There are different shapes, colours, and symbols or words. Text only signs **must not** be used.

There are **four** types of safety signs, and **fire** signs.

Prohibition

A red circular band and a cross bar.

Mandatory

A blue circle with a symbol inside.

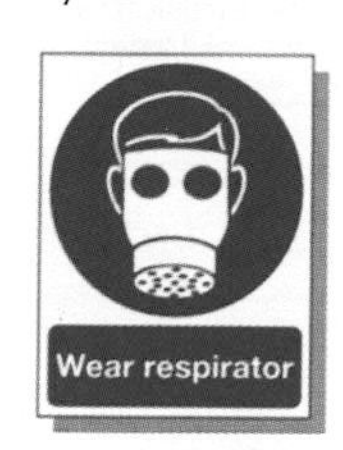

Warning

A yellow triangle with a thick black border.

Safe condition

A green square or rectangle with a symbol inside.

Fire equipment

A red square or rectangle with a symbol or text.

ACTIVITY

Identify the signs shown.

a ....

b ....

c ....

d ....

e ....

f ....

g ....

h ....

i ....

j ....

k ....

l ....

m ....

n ....

Emergencies

Emergencies may be caused by many things: a workshop accident, a fire, a spillage of a flammable or hazardous substance – even a bomb scare.

If you are sure there *is* an emergency, sound the alarm, evacuate the building and ring **999** for the emergency services. (See page 40.)

Fire alarm

In a small workshop, the fire alarm may be simply shouting '*Fire*'. In a large workshop an **automatic alarm** may be linked to the fire station, for immediate action.

If a workshop employs more than five people, it must have an emergency evacuation procedure. It must also have a building plan, and signs that show where to find:

- fire extinguishers
- fire exits
- assembly points
- first-aid points.

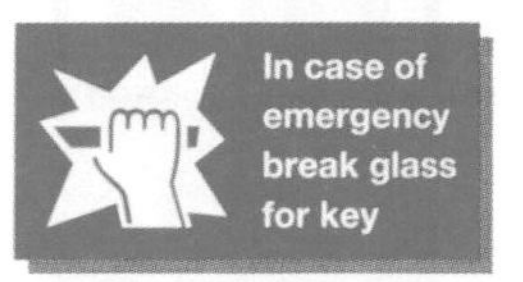

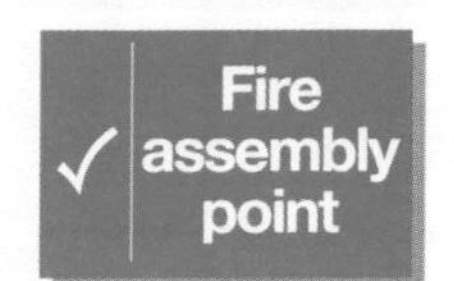

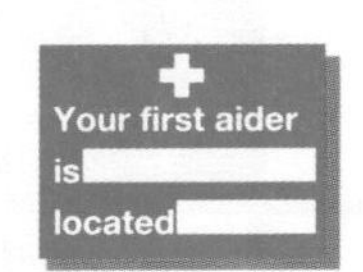

WORKSHOP

1. Sketch a plan of your garage workshop. Indicate the positions of
   - fire extinguishers
   - the fire alarm
   - machine isolators
   - first-aid equipment.
2. Write down your firm's evacuation procedure.

Keep this information in your portfolio.

Fire extinguishers

There are several types of fire extinguisher, suitable for different kinds of fire. Your garage should have extinguishers to fight fuel fires and electrical fires, as well as ordinary water extinguishers.

The table below shows types of fires, and the extinguishers recommended for fighting them.

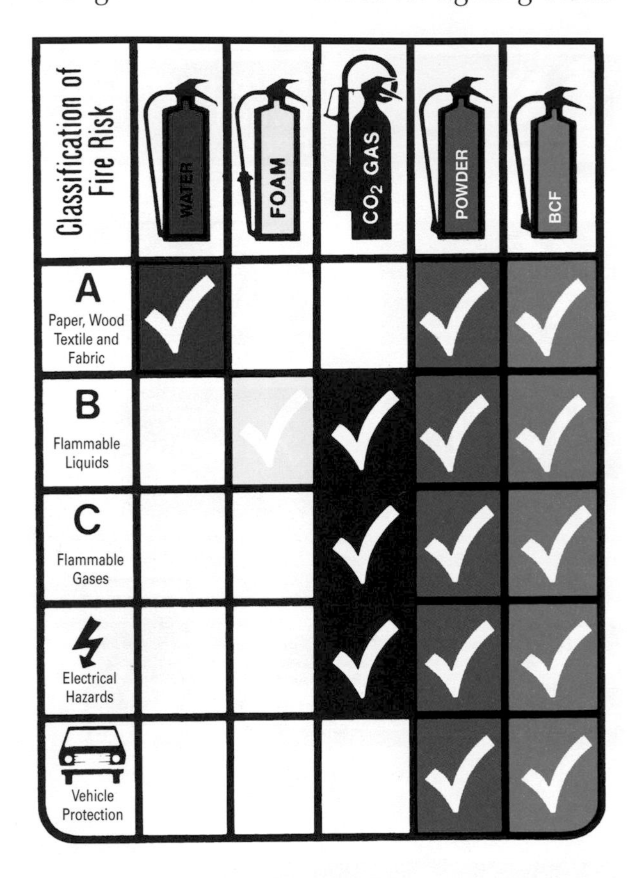

Classification of Fire Risk	WATER	FOAM	CO_2 GAS	POWDER	BCF
A Paper, Wood Textile and Fabric	✓			✓	✓
B Flammable Liquids		✓	✓	✓	✓
C Flammable Gases			✓	✓	✓
Electrical Hazards			✓	✓	✓
Vehicle Protection				✓	✓

WORKSHOP

1 Examine the fire extinguishers in your workshop. Note the types. Do they have a BS EN 3 kitemark?

2 Add this information to the plan of your workshop (page 19).

HEALTH & SAFETY

A fire blanket can be wrapped around someone who is burning. It smothers the fire.

Refillable Fire extinguishers

Industrial Carbon Dioxide Extinguisher

Industrial Water Extinguisher

New colours

Since January 1997 the British Standard for fire extinguishers has been **BS EN 3**. Under this standard, all fire extinguishers must be coloured *red*. However, 5% of the surface area may be colour-coded using the colours many people are already familiar with.

Extinguisher contents

The different kinds of extinguisher have different contents.

- **Water extinguisher** Water will kill the heat and put out the fire. This should be used for wood and paper fires. **Do not use it for electrical fires: you could get an electric shock. Do not use it for petrol or oil fires: burning fuel will float on the water.**
- **Foam extinguisher** The foam is water-based. It smothers the fire which goes out because there is no oxygen. This can be used with flammable liquids. **Do not use this in a garage.**
- **CO_2 extinguisher** This produces carbon dioxide gas, which removes the oxygen around the fire. However, because CO_2 does not remove the heat, wood and paper could re-ignite later.
- **Dry powder extinguisher** The powder is a fire-retardant dust. This covers the fire like a blanket.
- **BCF extinguisher** This smothers the fire with a blanket of heavy vapour. It is very clean, and leaves no deposit.

Good housekeeping

Maintaining a clean work area

We are all impressed when we see a clean and shiny car, even though we know it would work just as well dirty. In the same way customers will be impressed if you keep your workshop clean and tidy. No one wants to see a dirty workshop, a cluttered parts department, an untidy forecourt, or a patch of oil on the floor.

Remember – there are benefits in keeping the workshop clean:

- passing customers who need work done may be attracted in
- regular customers will be happy to keep coming back
- staff will work better
- accidents are less likely to happen
- work will be completed faster
- you are less likely to lose tools and parts.

Housekeeping routines

Each workshop will have a system to keep the place clean and tidy. As a new employee you may have to do some simple cleaning or tidying up as part of your duties.

A large company may employ cleaners; in a small one the work will be shared between the staff. All technicians are likely to be responsible for cleaning up after their work. Some jobs are particularly dirty – for example, jobs on exhausts or suspensions always leave dust and dirt on the floor.

Make sure you know your firm's housekeeping routines. Here is a typical daily housekeeping routine for a small garage workshop.

Opening

- Move parked vehicles away from the work areas.
- Check that the lifts and floor area are clean and free from obstructions.
- Check that all tools and equipment are clean and tidy.
- Check the reception counter is tidy, and that the sales computer equipment is working.
- Check that the reception area and seating are straight and tidy.
- Check the special workshop tools and equipment such as the air line, the tyre-remover and the wheel balancer are working.

During the day

- Keep the lift and the floor clean.
- After use, put special tools back where they are normally kept.
- Make sure that items such as wheels and removed tyres do not obstruct work areas or pathways.
- Make sure that the reception area is kept tidy, with ashtrays empty and magazines straight.

Closing

- Put away neatly all special tools.
- Lock up your personal toolbox.
- Check that any customers' vehicles are secure.
- Clean and tidy the work area.
- Switch off power to equipment.
- Tidy reception and empty the till.

ACTIVITY

List the daily and weekly housekeeping routines in your workshop.

Keep this list in your portfolio.

Delivery of goods

When goods are delivered, do not place or leave them where they would block walkways or exits.

HEALTH & SAFETY

Avoid personal injury. Do not lift anything too heavy for you – about 20 kg is a recommended amount.

If using lifting gear, never exceed the Safe Working Load (SWL).

Cleaning

Cleaning equipment should be kept in a separate store, as many chemicals are highly concentrated.

Always read the instructions on the labels before using them (see the COSHH regulations on page 16). If specialised cleaning is required, your employer will provide protective clothing.

When you have finished cleaning, put the cleaning equipment and unused chemicals back in the store.

HEALTH & SAFETY

While cleaning, place cones and notices to warn others. Section off areas that could be dangerous, such as slippery floors.

HEALTH & SAFETY

Some items may be hazards if broken or spilt. Watch out for caustic or corrosive chemicals, dented pressurised canisters, and aerosols.

Oily Waste Can

The self closing can protects the oily contents from fire outside the can and will snuff out any fire inside the can. The base of the can is raised to allow the flow of air and so reduce the risk of spontaneous combustion.

Emergency cleaning

Sometimes you will need to clean up after a breakage or spillage. Some oil might be spilt, for example, or some glass broken on the floor.

Breakages and spillages must be cleaned up immediately. If this is not done, someone may be injured. Also, the firm could be in breach of the Health and Safety at Work Act (see page 8). If an accident happened, the firm might be fined.

ACTIVITY

Name examples of cleaning substances in use where you work that have these symbols on their labels.

IRRITANT

1 ....

....

....

2 ....

....

....

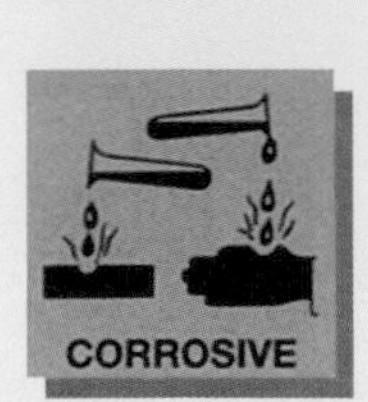

3 ....

....

....

4 ....

....

....

HEALTH & SAFETY

If you need to use a hazardous cleaning material, read the label on the container. This will tell you how to use it safely, and what to do if you *do* have an accident – for example, if the cleaning material touches your eyes or skin.

ACTIVITY

Make a list of the cleaning materials available in your garage. Think about:

- *personal cleanliness* – e.g. barrier cream, hand-cleaning material
- *equipment* – e.g. lifts, trolley jacks, wheel-balancing machines, special and personal tools
- *the workshop* – e.g. the floor area, car bays, walls, windows, lighting.

Keep the completed list in your portfolio.

HEALTH & SAFETY

Before cleaning or removing guards from an electrically powered machine, make sure it is electrically isolated.

Find out where the isolators or stop buttons are.

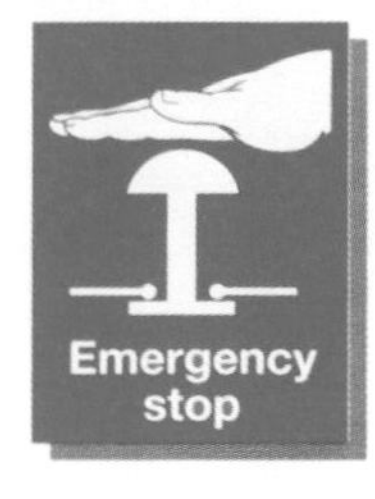

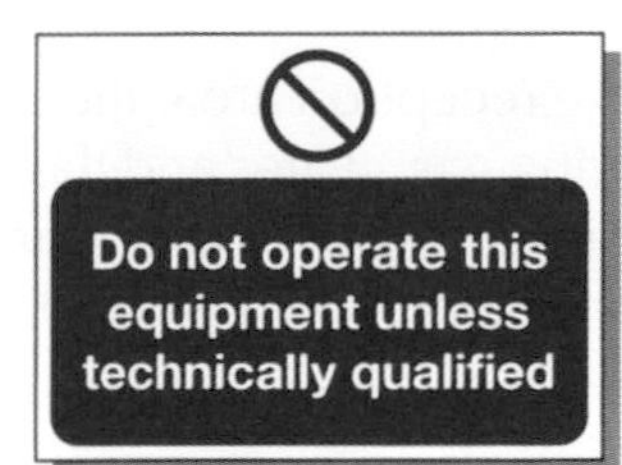

Disposing of dangerous waste material

All workshops produce **dangerous waste** materials – dirty engine oil and filters, scrap, tyres, exhausts and batteries, broken or scrap plastic or metal components, and waste paper.

Items must be disposed of in different ways. Usually this is decided by the local council, who pass **by-laws**. Refuse disposal requirements differ from place to place.

Some types of dangerous material must be kept separate. They will be collected by specialist agencies, or taken to the local refuse collection point. All must take note of the Environmental Protection Act.

ACTIVITY

Tick the types of waste produced by your workshop. Alongside, state the correct method of disposal.

Type of waste		Method of disposal
Engine oil	☐	
Oil and fuel filters	☐	
Brake fluid	☐	
Batteries	☐	
Tyres	☐	
Exhausts	☐	
Metal scrap	☐	
Plastic scrap	☐	
Paper and packaging	☐	
Household waste	☐	

Workshop resources

Every workshop has many **resources**. You need to be aware of what they are, and how you can make best use of them. Resources include:

- *special workshop tools and equipment fixtures*, such as vehicle lifts, the air compressor, the wheel-balancer, steering alignment tools, the headlamp aligner, and so on
- *stock* in the stores – different types of tyres, exhausts, shock absorbers, batteries, oil, and so on
- *fixtures and fittings* in the reception area, the staff dining area, and elsewhere
- *utilities*, such as electricity, gas, water and the telephone
- *space* available – to work, to store parts, to park and display vehicles
- *time* – often the firm's most valuable resource.

Use resources to the best advantage

Tools

Use tools and equipment safely and properly. Avoid damage, and don't risk your own health and safety, or anyone else's.

Utilities

Electricity, gas and telephone calls are costly. Waste will reduce the firm's profit. To save energy, your firm will probably have some automatic timing controls fitted.

Consumable items

Do not waste consumable items, even if they are small. After fitting an exhaust system, for instance, return to stock any unused components such as nuts, brackets or rubber mounting rings.

Security

Make sure that parts are kept as safe as possible. If visitors wander around the workshop and are not observed, they could steal things. Theft by staff may also occur.

Taking goods, equipment or money without permission is always **theft**. Theft by staff is **gross misconduct**, and can lead to dismissal.

Do not leave keys in unattended cars. It is not uncommon to have cars driven away from the forecourt.

TIP
If you see someone acting suspiciously, ask them what they are doing. If the answer is not satisfactory, tell a senior member of staff immediately. **Do not put yourself at risk.**

If the person runs away and you suspect they have stolen something, you may wish to tell the police. You may then be asked to say what took place:

- the date and time of the incident
- what took place
- a description of the person.

Using resources economically

Avoid wasting power

- Turn off lights when they are not needed.
- Keep workshop doors shut, to keep heat in.
- Report faulty components – for example, a leaking air line that causes the compressor to keep switching on.
- Turn off water when you are not using it, especially when washing cars.

Use space sensibly

Workshop space is expensive. Costs include rates, taxes, heating and lighting. Do not *waste* this space!

- Place vehicles so that you can work properly, but so that they take up minimum space.
- Keep the working space around ramps and in gangways clear of obstructions.
- Clear away quickly when a job is finished.

Time

You rightly expect to be paid for your time at work. So you must play your part and help the firm to make a profit.

- As a newcomer to the trade you are unlikely to work as quickly as a skilled technician. But aim to build up your speed.
- Work steadily, but not slowly. Time wasted is almost impossible to make up. You may have a bonus at stake!
- Organise yourself. Gather tools, information and equipment before you start a job.
- Complete jobs properly. Putting things right later is costly.

TIP
Don't waste time. When you have finished your job, ask for something else to do.

Clear away waste items. Make the place safe and tidy.

QUESTIONS

1 Suppose you take six minutes extra at break-time, morning and afternoon. How much does this add up to in:

a one day?................................

b one week (5 days)?.................

c one year (52 weeks)?..............

2 Is the total more ☐ or less ☐ than one week's work?

Keep down telephone costs

- Use cheap times for non-essential calls.
- Obtain permission for personal calls. (Without it, they could be classed as theft.)
- Decide what you are going to say before making a call.
- Keep calls brief.

ACTIVITY

Find out the costs of local and national calls. When are they cheapest?

Chapter 2

Working relationships

In this chapter you will learn about:

- **working with others** – how to develop good relationships with colleagues
- **working in a team** – knowing your own job, and what other people do

To be successful at work, you need to know your job. But technical skills are not enough. To be happy doing your job you also need to get on with the people you work with, and with customers.

This chapter gives some simple suggestions to help you work well with other people.

Working relationships

There are several different kinds of **relationships**. You have a *family relationship* with a relative – a parent, a child, an aunt or a cousin. With people you like at school or college, you have a *friendly relationship*. You may have a *romantic relationship* with a girlfriend or boyfriend.

Working relationships develop with the people who work alongside you. As you interact in the workshop with the manager, other mechanics and fitters, or your supervisor, you build working relationships with them.

Good working relationships are very important to the success of every business. You may not be *friends* with all your colleagues; occasionally you may even *dislike* some of them. But to help the business run smoothly, you must get on with all of them professionally.

The company's aim

It is important to be aware of the *aim* of the company for which you work. This may be set out in what is known as a **mission statement**. This explains the main objectives of the company: the reasons why it is trading, and what it hopes to achieve.

All firms are in business to make a **profit**. This is shared between the owners or any shareholders. A profitable business is likely to be a successful one, and a successful business can offer its workforce **job security**.

QUESTION

What benefits might flow from a profitable business?

1 ....

2 ....

3 ....

Working as a team

To make the company successful, all of its employees must work together. They must **co-operate**, like members of a football team: this is **teamwork**.

If the firm does well, and employees get on with each other and trust each other, there will be a good feeling in the workplace, and people will be enthusiastic about their jobs. This feeling is called **good morale**.

When everyone works hard, and no one wastes time or resources, the firm will be **efficient**. By being efficient, employees will get a lot done – they will be **productive**.

If a team with good morale works efficiently and productively, customers will be satisfied and pleased to come again. They are also likely to recommend the firm to others, so it will gain a good **reputation**. And this in turn will bring more business, and the firm will become even more successful, and will grow. It will gain a good **company image**.

Building good relationships

Here are some ways in which you can build good working relationships with your colleagues.

- Recognise that there are differences in personality and temperament.
- Treat colleagues politely and with respect.
- Co-operate, and assist willingly with requests.
- Talk with colleagues about problems, changes or proposals.
- If something you are doing goes wrong, or if you break something, tell your supervisor straight away.
- When a working relationship breaks down, be honest and fair, and try to put things right.

Here are a few examples of the kind of things that can upset good working relationships.

- In everyday conversation you may discuss your social life, sports, films and so on. You may find that a workmate has opinions on some topics which are different from yours. Never allow these differences to spoil your working relationship.

QUESTION

Name three *other* topics which could create argument and bad feeling.

1 ....

2 ....

3 ....

- Sometimes you may find colleagues who lack interest or enthusiasm, who are lazy or incompetent, who keep being absent, and so on. If people do not 'pull their weight' in the workplace, this can cause anger and frustration among other members of the team.
- Beware of anyone who ignores company rules and regulations, or safe working practices. This behaviour can create problems or even dangers for everyone else in the company.
- Personal appearance and hygiene are important. Employees who do not bother with these may upset colleagues, and this may affect the performance of a workforce.
- Managers and supervisors should not show favouritism and should pay everyone fairly.

The company image depends in part on the way its workers are judged by customers. Customers will see your work, and also how you behave and interact with your colleagues.

ACTIVITY

List three actions by employees that might spoil customers' opinion of the firm. Could these actions be avoided?

1 ....

2 ....

3 ....

Organisational structure

The **organisational structure** describes who does what in the company. It describes each person's duties and level of authority. It also shows everyone's working relationships, from the managing director to the most junior employee.

Job description

A **job description** names a job title and states the duties and responsibilities of a particular job.

Each employee should be aware of their own role and the roles of others. Disputes may arise if anyone is uncertain about who does what.

An organisation chart such as the one below shows levels of authority and responsibilities. **Vertical links** show the line of authority from senior staff downwards. **Horizontal links** indicate people with equal status and authority in the company.

ACTIVITY

Write out your own job description.

JOB DESCRIPTION
JOB TITLE

....

DUTIES

....

....

....

....

....

Responsible to:

....

....

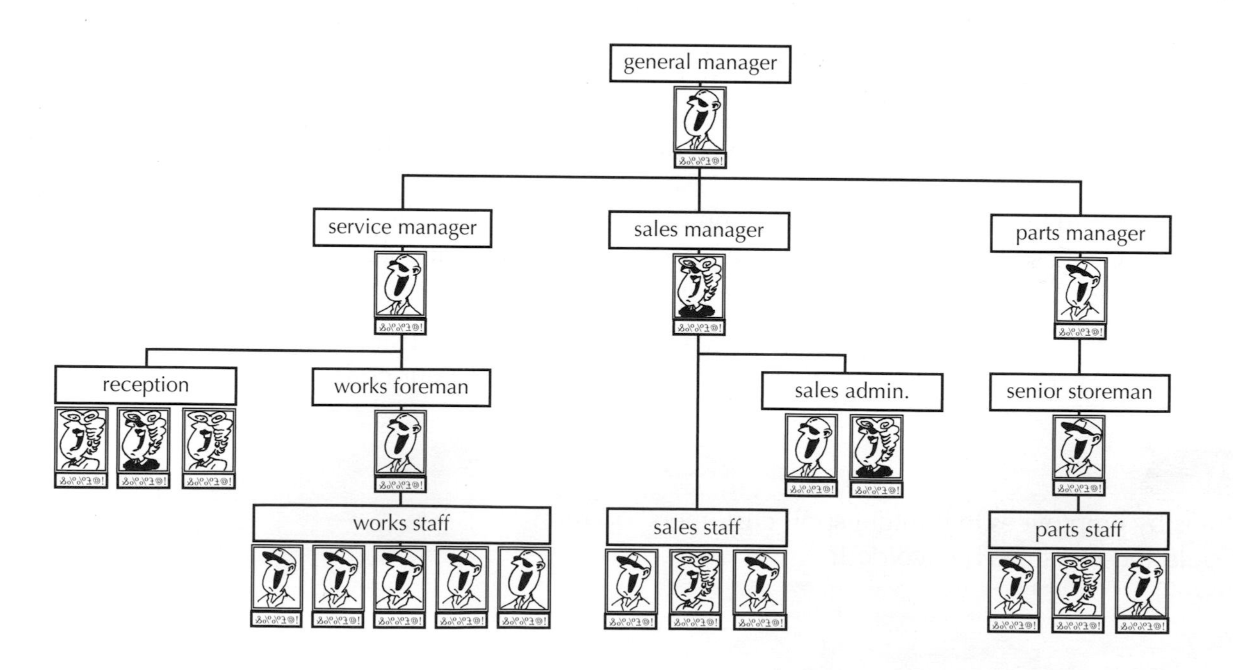

ACTIVITY

Draw a chart to show who is in charge of each section, either where you work or where you go to college.

Place the completed chart in your portfolio.

Communication

When problems and misunderstandings do arise, it is often because of poor communication. Good **communication** is an essential part of every working day, for employees at all levels.

In a typical garage, there will be daily communication between colleagues within a department, between departments, between management, supervisors and shop-floor staff, and between customers and reception staff. The company will also communicate with suppliers, subcontractors, vehicle manufacturers, advertising agencies, banks, accountants, lawyers, the council, and so on.

ACTIVITY

Suggest three *methods* of communication which might be used.

1 ....

2 ....

3 ....

Non-verbal communication

It is not only your words that communicate – so too do your facial expression and the way you stand and move. Customers will notice your **body language**: for example whether you smile and look directly at them or slouch and avoid their eyes.

ACTIVITY

This section is particularly concerned with the *effects* of communication on working relationships. The method of communication chosen will depend on the situation.

For question **1** below we have provided examples of situations where you would talk to someone face-to-face. For questions **2** and **3** you need to give examples of situations where you would *write* to someone or *telephone* someone.

1 *Direct discussion* – face-to-face conversation.

- **a** With a customer, to find out their requirements and to advise them.
- **b** Between colleagues, about a particular job.
- **c** With management and supervisory staff, to talk over procedures and problems.

2 *Writing*

- **a** ....
  ....
- **b** ....
  ....
- **c** ....
  ....

3 *Telephone*

- **a** ....
  ....
- **b** ....
  ....
- **c** ....
  ....

Problems in communication

Some things can cause problems in communication.

Lack of communication

Problems can arise if people are not told things they need to know. Changes to systems, safety issues, shop-floor problems and the like must be passed on to the right people without delay.

Delays in getting information to staff may be caused by pressure of work or different hours of work (such as part-time, full-time, or shift work).

Incorrect communication

Problems can also arise if information given is wrong. A good example of this is when a spoken message is passed on and wrongly remembered – always write down messages for people, while the information is clear in your mind. Similarly, if a job sheet lacks some detail this causes confusion.

TIP
Each company should have a *grievance and disciplinary procedure*. Find out about yours.
Place a copy of any printed information in your portfolio.

Good communication

Chapter 3

Communication

In this chapter you will learn about:

- **the telephone** – listening properly and taking messages
- **customers** – how to speak to them and deal with queries
- **written messages** – different means of communication within the workshop and beyond
- **goodwill** – how to build it and keep it

If a workshop is to run smoothly, all of the staff must be able to communicate with one another and with the customers.

This chapter shows a number of ways of communicating, and how to use them to get the best results.

The telephone

> **TIP**
> When answering the telephone, always have pen and paper ready to take messages.

The telephone is a vital part of running a business. You need to use it well, and efficiently.

At home you may be used to chatting on the telephone, even to someone you will see later that day. You can be relaxed and take your time. In a garage reception centre, though, a call that takes longer than it needs to wastes money and may stop another caller getting through. You might lose business! In a small garage there may be only one telephone, at the office or reception area. The foreman may carry a portable telephone. At a main dealer, a telephone switchboard is used: incoming calls are answered by a switchboard operator who then transfers the call to the right person.

Remember: how you answer the telephone may affect how the garage is judged by the caller. If you welcome callers promptly you will create a good image. The way you speak and the tone of your voice are important too. When you answer the telephone, try to speak in a friendly, cheerful way, so the caller feels that you are interested and really want to help.

Answering the telephone

A 10-point checklist:

1. Answer, within 2 to 8 rings.
2. Speak slowly and clearly.
3. State the firm's name, and give your own first name.
4. Ask how you can help the caller.
5. Listen to what the caller says.
6. Write down the caller's name and telephone number.
7. If there is a message, write it down.
8. Complete the call by thanking the caller.
9. Replace the telephone receiver correctly, so the next caller can get through.
10. Keep calls brief and efficient.

Use the rule: *pick up – fix up – shut up.*

Types of calls

Incoming calls are likely to be one of these types:

- calls to be transferred to another employee
- queries
- messages from customers
- personal messages for staff
- wrong numbers.

> **TIP**
> When answering the telephone, smile – the caller will be able to hear the friendliness of your voice.
>
> 'Good morning, New Day Garage. This is Chris speaking, how may I help you?'

Listening skills

Are you a good listener? Simply *hearing* is not the same as **listening**. When someone speaks to you on the telephone, concentrate on what the caller is saying. Don't just think about what you are going to say in reply, and don't butt in. Ignore what is going on around you. Never try to talk to someone else at the same time.

Taking messages

Always *write down* messages, however brief. Repeat the message you have recorded, so that the caller can check that you have got it right.

TIP
Before asking a caller to hold on, make sure that she or he is *willing* to wait.

Telephone message sheet

Your firm may use printed message sheets. These should include:

- the name of the caller
- who the message is for
- who the message is from
- the caller's telephone number
- the date and time of the call
- the *urgency* of the message
- the message itself
- how the person called should respond – for example, call back or wait for the caller to call again.

Telephone Message

Message for ______ Dept ______
Date ______ Time ______
From ______ Company ______
Phone No. ______

☐ Urgent ☐ Please ring ☐ Will call again

Message

Taken by ______

ACTIVITY

Complete at least six message sheets while taking telephone messages. Keep them in your portfolio.

QUESTIONS

1 What is wrong with the following messages?

To Jane

Someone rang about the cost of fitting an exchange gearbox. Please ring him after 11 today.

To Sanjay

A man said he would bring his car in tomorrow.

Ricky

The man you saw last week is willing to make an offer for the Ford in the showroom if you give him a ring.

a Jane's message

....

....

b Sanjay's message

....

....

c Ricky's message

....

....

2 What problems could arise for the people who received these messages?

....

Transferring calls

You will not be able to answer all queries yourself. Instead you will need to find someone who can. You need to know where each person works, so you can transfer calls when necessary.

If there is a delay, take the caller's number and say that you will ring back. Give a time limit, and call within that time even if you have not got the answer to the query.

TIP
Don't leave the telephone caller on the line for more than a few seconds. Let them know why there is a delay.
Being left on the end of a line can cause 'telephone rage' and the caller may be offensive or hang up.

ACTIVITY

Name the departments within your firm. Name the people within these departments who can give information. What are their telephone numbers? Place the answers in your portfolio.

Urgent messages

When you receive an urgent call for someone at work, it is up to you to make sure he or she gets the message.

Be sure to write down the name of the person calling, the telephone number, the person the message is for, the message itself, and any other helpful information.

Confidentiality

Some information is **confidential** – you must not give it over the telephone. Examples include employees' addresses and home telephone numbers, and vehicles' trade purchase prices.

ACTIVITY

What other information must be kept confidential?

1 ....

....

2 ....

....

Outgoing calls

You may or may not be allowed to make outgoing calls. If you are, all the same issues will arise as when taking incoming calls.

Unless the call you have to make is simple, write down the telephone number and the message before you start.

Telephone directories

Your firm should have a local British Telecom (BT) directory, ***The Phone Book***, a local ***Yellow Pages***, and maybe a ***Thomson Local***. Spend some time looking at these books.

At the front of *The Phone Book* are 'Help and Advice', followed by 'Business and Services' and then 'Residential' numbers. Also listed are the UK area codes and international codes.

The *Yellow Pages* include street maps for local towns and a directory of *types* of businesses, in alphabetical order.

The *Thomson Local* is a business directory. Names are in alphabetical order; and then there are yellow pages which show the *types* of local businesses, again in alphabetical order.

ACTIVITY

Use the telephone directories to find:

1 An alternative number for 999. ....

2 The BT fault reporting number for business customers.

....

3 The UK directory enquiries number.

....

4 The call return service, which gives you the number of the person who last called.

....

5 The number to press on your telephone to redial an engaged number.

....

6 The UK area code for Aviemore.

....

7 The international code for France.

....

8 A local number for a *national* exhaust/tyre repair firm.

....

9 The number of a *local* exhaust/tyre repair firm.

....

10 The nearest main Ford dealer's number.

....

Emergencies

In an emergency, for **fire**, **police** or **accident**, press or dial **999**.

1 Tell the operator which service you want.
2 Wait for the operator to connect you.
3 Tell the emergency service:
   a Where the trouble is
   b What the trouble is
   c Where you are
   d The number of the telephone you are using.
4 Let the person at the other end of the telephone ask the questions.
5 Don't hang up until the service has all the information it needs.

WARNING

Never make a false emergency call. It is against the law, and you could risk the lives of others who *really* need help.

You can be traced immediately to the telephone where the call came from.

Oral messages

Messages given by word of mouth (not written down) are called **oral communications**. Misunderstandings can arise.

You should listen carefully to what you are told. At work you will often be told what jobs to do. Unless you listen properly, you might do the wrong job, or even work on the wrong car!

If you are told to do a job or given a message that you do not understand, ask questions until you do.

When you give a message to someone else, ask this person to repeat the message. You can then check that you have made yourself clear.

TIPS

- Be an active listener.
- Look at the person who is talking to you.
- Nod and smile to show understanding. Don't be afraid to ask questions if you *don't* understand.

Relaying messages

In small firms many messages will be passed by word of mouth. This is not always reliable. For anything important, write it down to make sure the details are correct and you don't forget the message.

On the telephone, oral messages can be misheard.

TIP

To get customers to discuss what they want, try to ask questions that include such words as *who, what, why, where, when* or *how*.

You will get more information this way than by asking only questions that require *yes* or *no*.

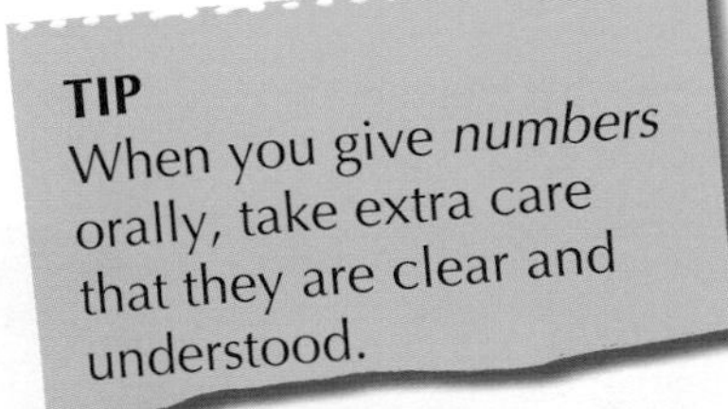

There are 70 cars prepared for sale outside on the forecourt, 19 are blue.

There are 70 cars prepared for sale outside on the forecourt and 9 are blue.

There are 17 cars prepared for sale outside on the front and 9 blue.

There are 17 cars prepared for sale outside and 9 black.

There are 17 cars prepared for sale outside and 9 bikes.

If a message is passed orally via several people, it may become totally changed!

Speaking to customers

When speaking to customers – face-to-face, or over the telephone – try to vary your tone of voice. A single tone of voice will give the impression that you are not interested, and it can be very boring.

Your style of talking will vary, depending on who you are talking to. The style and slang you might use when talking to your friends must not be used when talking to customers. Use a formal, polite style. Avoid technical language unless the customer requests it.

Remember: if customers are dissatisfied, they will take their work elsewhere.

TIP
Remember: it is part of your job to keep the customer happy, so that she or he goes on using your firm.

The customer brings work and so pays bills. These bills pay your wages.

ACTIVITY

Body language is one kind of communication. How can *your* body language show that you are interested and alert?

Imagine someone listening to *you*, with real interest. Describe this person's

1 eyes ....

....

2 facial expression ....

....

3 head gestures ....

....

Written communications

Messages within the company

Large garages will need to pass written messages between departments. These are known as **memoranda** or **memos**.

ACTIVITY

1 Write a memo to Dave in the parts department. Ask him whether he has in stock the front suspension coil springs for the car you are working on. If not, how soon can he get them? Keep the memo in your portfolio.

2 Look at the memo opposite. Find out what *re* and *cc* mean.

NEWDAY GARAGE

MEMORANDUM	
From:	To:
Date:	

Re:

CC:

Messages outside the company

Messages can be sent to other firms or customers in various ways. A small firm will probably use only the first two or three of the examples below; a large organisation may use all five.

Post

Post can be used if the message is not urgent. This is suitable for a customer's bill or a service reminder.

Telephone

The **telephone** gives immediate contact, so messages and actions can be quickly dealt with.

Fax

A **fax** is a written message or drawing, passed by telephone line. This allows information, including pictures, to be transferred very quickly.

E-mail

An **e-mail** message is sent electronically, via the telephone system, from computer to computer.

Internet

The Internet is a widely accepted, speedy, inexpensive and efficient means of communication. It may be used to obtain information on almost any topic, to advertise goods and services and conduct many business transactions such as ordering stock from manufacturers and suppliers. As with email, which travels via the internet, the system operates via telephone lines and an ISP (Internet Service Provider) from computer to computer anywhere in the world.

Building goodwill

Trust and **goodwill** with customers will be very much helped if you practise good, clear communication. Misunderstandings can easily make customers go elsewhere. In Britain, customers who are not satisfied tend not to complain – they just don't come back. So you may never find out that something is wrong – until, perhaps, there is no work!

Keep customers informed

To help keep goodwill, contact the customer if:

- a vehicle repair cannot be completed in time
- extra work is needed
- there is a delay in obtaining parts
- the appointment time needs changing.

Failure to do these things might seriously annoy the customer.

Promote a good company image

What do customers want to see?

- helpful, friendly staff, working in a clean and tidy reception area and workshop
- organised, efficient staff who know what they are doing
- staff who know about the vehicles they work on.

How should you appear at work?

- clean, tidy, and dressed in company overalls
- organised
- ready and willing to help.

ACTIVITY

1 Suggest three things you personally can do to improve or maintain customer goodwill.

a ....

....

b ....

....

c ....

....

2 List four things that you think customers want from a repair shop.

a ....

....

b ....

....

c ....

....

d ....

....

Chapter 4

Handling and merchandising stock

In this chapter you will learn about:

- **new stock** – receiving and unpacking it
- **storing stock** – safe storage and stock rotation
- **displaying and merchandising stock** – planning and labelling, and sales law

To work efficiently and to keep customers happy, you need to have the right goods in stock. They must all be handled on delivery, identified, labelled, and stored securely for future use. As with all jobs, some knowledge and skill is needed to get the best results.

In selling goods – especially accessories, which often show a high profit – an attractive display can work wonders. This chapter has some useful ideas, along with reminders of the relevant law.

Handling stock

Receiving stock

Servicing, repairing and selling motor vehicles involves replacing worn-out or faulty **parts**, and selling items such as vehicle **accessories**. Ordering and receiving stock is therefore part of the daily routine.

Some goods received will be for use in the service and repair workshop; others will be supplied 'over the counter' to customers. The parts that make up a motor vehicle vary considerably in value, size, shape, weight, and how easily they are damaged.

Most workshops or parts departments will have an area where goods are received. The main activities there are receiving new stock, unpacking, dealing with problems such as damaged stock, and placing the stock in the correct locations.

When incoming stock arrives, it is important to check it.

- Check the delivery against the **delivery note**. (This document comes with the goods.)
- Check the delivery against the **order**.

At this stage the goods must be *identified, counted* and *checked for damage.*

When a large quantity of stock arrives, it is not practical to unpack and check the complete delivery before signing for it. Normally, therefore, a signature acknowledges *receipt* of a delivery *without checking*. A thorough check is then carried out as soon as possible. Nevertheless, it is important to examine the outside of crates and packages for any obvious damage, while the delivery driver is still there.

ACTIVITY

Complete this sample document to show a typical delivery note.

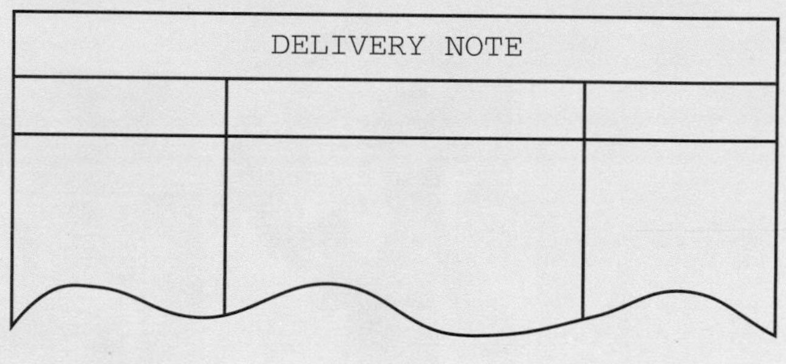

Safe handling

Make sure you know the correct lifting and carrying techniques, and how to use lifting equipment safely.

Protection of goods

Check the markings on the containers. Take particular care with vulnerable parts, such as glass and expensive electronic components.

TIP
The document supplied with incoming stock maybe called an **advice note** or **consignment note**. With imported goods, it may be a **bill of lading**. Sometimes it is simply a copy of the invoice, but without prices.

ACTIVITY

Find out what these symbols mean.

1 .... 2 .... 3 ....

Stock security

Do not leave stock unattended if the outer doors are open. Keep the stockroom locked, and don't leave keys lying around. Follow your company's security procedures.

ACTIVITY

1 Circle the sketch that shows the correct way to lift.

a **b** **c** **d**

2 Name the risk indicated by each sign.

a .... **b** .... **c** ....

Unpacking stock

If the activities of receiving and unpacking goods are to run smoothly, the **procedure** must be organised.

Keep the work and storage area clean and tidy. Make sure there is a definite place for everything connected with the job. This includes handling equipment, tools for unpacking, waste packaging materials, stock to be returned, and so on. Keep things in their proper places.

To cut through adhesive tape on boxes, a **Stanley knife** is useful. Take care not to damage the contents, and not to injure yourself.

QUESTION

Name three other tools you might use for unpacking goods.

1 ....

2 ....

3 ....

Packaging

Goods in transit are carefully packaged, for a number of reasons: security, ease of handling, safety, and to protect the goods.

The way in which goods are packed depends on how easily they might be damaged, their weight, and how hazardous they may be.

QUESTIONS

How would you expect the following items to be packed?

1 Engine..........

2 Radio..........

3 Headlamp unit..........

Labelling

It is usual to identify all motor vehicle parts and materials with a **label**. For example, without the painted labels on their containers, how would you know which container held engine oil and which held automatic transmission oil?

QUESTIONS

What *type* of labels would you be likely to find on the following items?

1 New tyre..........

2 Silencer..........

3 New battery..........

4 Window glass..........

Disposing of packaging materials

Packaging materials are normally separated for disposal.

- **Pallets** and **crates** are usually returned to the supplier.
- **Paper** is binned, ready for the waste paper collection.
- **Contaminated materials** are stored safely and collected later by a licensed disposal company.
- **Recycling** – follow your firm's policy carefully.

Remember: you must recognise and keep together containers, crates and pallets to be returned to the supplier. If these are lost or damaged, your company will be charged for them.

ACTIVITY

Find out which goods are delivered to your workshop on pallets, in containers or in cages. Note whether exchange or warranty parts have to be returned in the same containers, or whether the containers should be returned empty or scrapped.

Place this information in your portfolio.

HEALTH & SAFETY

1 All areas must be well lit. Replace light bulbs immediately.
2 Keep the floor area clean and uncluttered. Clear spillages without delay.
3 Know fire-fighting procedures: fire drill, and how to use the fire extinguishers.
4 If any liquid is spilt from damaged containers, approach it with caution. You need to find out what it is before you can clear it up safely.

Dealing with discrepancies

A **discrepancy** in a stock delivery could be any of these problems:

- **Incorrect quantities** Check the quantity actually *delivered* against the quantity on the *delivery note*, and the quantity delivered against the quantity *ordered*.
- **Incorrect goods** Check the part numbers and descriptions: the delivery note should match your company's order.

- **Damaged goods** Some damage may be acceptable, such as a slightly dented can of oil. However, a similar dent on a new radiator grille would not be acceptable. If there is any doubt, ask your supervisor.

The delivery system

In a typical system the supplier has a national **parts distribution centre.** Dealers order parts using a computer network. Stock is delivered during the night, to ensure that supplies are ready for the next day's work.

Parts are supplied in a cage, with the delivery note. The driver wheels the cage into the delivery area, and collects the cage from the *previous* delivery.

The cage being returned may not be empty. It will contain any parts that are being returned for replacement, together with a **parts claim form**.

This system can only work when there are trust and full co-operation between the dealer and the supplier. If the dealer fails to return a cage, the supplier may fine them.

Depositing stock

Once stock has been identified and checked, it should be moved to the correct selection or storage locations.

Where a particular item of stock is located, and on what type of storage equipment, depends on the item's size, shape, weight, and hazard risk. **Bin units** and adaptable **racking systems** are widely used for motor vehicle parts storage.

Before you place stock in the store, be sure that you have found the correct location. Every part has a **part number**. This can be linked to its **bin location** within the store.

Each row of bins may be numbered, each bin unit lettered, and each shelf numbered. This reference would be shown on the computer against the part number: for example

Part no.: 77 00 850 292
Bin location: 4/B/7

TIP
Damp and high temperatures may affect some goods. Take these into account when placing items in the store.

Stock rotation

When you shop in a supermarket, you may notice new stock being placed at the *back* of the shelf. This ensures that older stock is sold first. This system of **stock rotation** is used with vehicle parts too.

QUESTIONS

1 What do the letters 'FIFO' stand for?

....

2 The components and systems on a vehicle can be shown in 'exploded view' on microfilm, and displayed on a screen. What is this system called?

....

3 Certain containers or packages display a 'HAZCHEM' notice. What is this?

....

....

....

4 Where would you see the letters 'SWL'? What do they mean?

....

....

....

5 If a substance is marked 'TOXIC', what hazard does this present to you?

....

....

HEALTH & SAFETY

As you move around the store, placing stock, be aware of the movements of other workers. If you place stock temporarily on the floor, for instance in an aisle, place warning signs near it.

Tyre racks

Tyres are normally stored on racks. As you can see in the picture opposite, tyres are stored with the labels to the front. This makes it easier to identify and select the tyre you need.

Tyres should be stored upright and in the shade. As with other components, they are placed on the racks and selected for use according to the FIFO system.

Displaying and merchandising stock

The **point of sale** is the place where customers part with their money – the sales counter. It is at this point that customers may be tempted to buy extra goods – more than they came in for.

One method of promoting extra sales is the **display**. Merchandising is all the activity related to selling the goods once they have reached the point of sale – layout, display, promotions, pricing, and so on. The effectiveness of the display in attracting customers and persuading them to buy is called its **customer impact**.

Goods can be displayed near to the point of sale, or in an area through which the customer passes on the way to and from the sales counter, such as the car showroom if this is on the way to service reception.

Planning a display

Here are some of the things you need to think about when planning and setting up a display:

- what stock is to be displayed
- the location and size of the display
- the layout and design of the display
- the impact on the customer
- the display equipment (stands, shelves, racks and lighting)
- signs and labelling
- the upkeep of the display
- security
- safety.

The **display equipment** required depends on what items are to be displayed. It must be strong and secure enough to support the display safely.

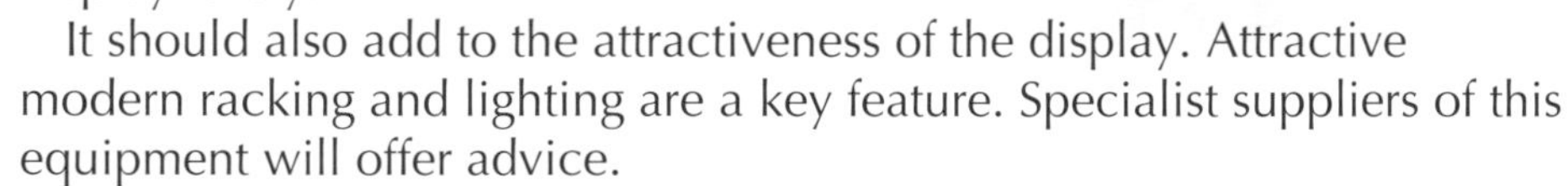

It should also add to the attractiveness of the display. Attractive modern racking and lighting are a key feature. Specialist suppliers of this equipment will offer advice.

Vehicle manufacturers will supply specially designed display stands for certain items, such as wheel racks and accessory cabinets.

Remember that customers respond best to goods displayed at eye level.

What to display

Even in a sales outlet with only a limited range of products, such as a fast-fit battery shop, the display can still show the range of goods and services available. In a motor vehicle parts store, though, the display can be more extensive. For example, it could feature:

- *Accessories* Road wheels, floor mats, steering wheels, in-car entertainment.
- *Special offers* Repair kits, clutch and braking system components, servicing parts, valeting products.
- *Seasonal items.*

ACTIVITY

Give some examples of seasonal items.

....

....

....

Designing a display

Effective **design** and **layout** of a display depend on specialised skills. However, you can follow certain guidelines when setting up your own display.

- Learn from the experts – look at other displays, in stores, garages, and elsewhere.
- Group similar items together.
- Do not try to display too much – keep it simple.
- Link certain parts of the display – develop a theme, such as 'servicing requirements'.
- Use colour and lighting to create impact.
- Make use of manufacturers' supply material and ideas for displays.

TIP
For a display to maintain its impact, it must be kept clean and in good condition, and it must be changed frequently.

Labelling

Signs and labels should be clear and simple, with not too many words. Signs should provide sufficient information about the products.

If you have to label anything,

- remove old, outdated labels
- check that the information is correct
- use the appropriate type of label. Word-processing and graphics pages on the computer enable you to produce very professional looking labels.
- make sure that the label is secure, and easy to see and read.

HEALTH & SAFETY

Return excess stock, materials, tools and equipment to their correct locations. As always, keep the display area clean and tidy and free from hazards.

ACTIVITY

Make out a sales label for a display item – choose a motor vehicle accessory. Include:

- the name of the item
- information about it
- any promotional information
- the price.

Law relating to sales

Trade Descriptions Acts 1968 and 1972

The **Trade Descriptions Acts** deal with *false* or *misleading* descriptions of goods or services. When you are involved in selling goods or services you must not:

- describe products falsely
- give misleading information about your products or services.

If you are setting up a sales promotion on a display you must not:

- offer products at a 'sale price' unless they have previously been on sale at the full price for at least 28 days
- make false or misleading price comparisons.

Sale of Goods Acts 1979 and 1994

Goods offered for sale must be:

- of satisfactory quality – they must be suitable for the purpose for which they are supplied
- fit for the purpose – they must do what is claimed for them
- as described – they must fit the description.

Chapter 5

Tyres

In this chapter you will learn about:

- **tyres** – the many different types, and how they are constructed
- **valves** – tyre inflation and tyre pressures
- **wheels** – repairs and balancing
- **tyre repairs** – tube repairs, remoulds and retreads
- **specialist tyres** – for heavy commercial vehicles, agricultural vehicles, earth-movers and industrial vehicles

Tyres may not seem very glamorous, but no car or lorry can run without them. What is more, unless they are in good condition and of the right sort, the driver might well break the law. And as a car mechanic, so might you, if you don't do your job properly. But don't worry, this chapter should make sure you don't end up behind bars!

Tyres are very expensive. This chapter explains about the different types of tyres, how to look after and repair tyres, and how to recognise various faults with tyres. The chapter also explains about wheels and wheel balancing.

There are safety aspects that you must watch for, and points of law that you must know. These affect not just cars but commercial vehicles, agricultural vehicles, earth-movers, and industrial vehicles.

Tyre types and use

A tyre may look like a large black rubber doughnut, but in fact it is a very cleverly designed piece of technology. A lot of skill goes into the making of each tyre. And because of this, tyres are expensive, as you will know if you have ever had to buy one.

If you follow racing or rallying, you will know how important it is to choose the right tyres. It is *so* important that it is always the driver who has the final say about which tyres to use. This is because it is *through* the tyres that all the vehicle's forces meet the road surface. The car must have the right tyres so that the driver can accelerate, brake and steer safely. The tyres also support the entire weight of the vehicle and its contents: the right tyres will help give a comfortable ride.

The types of tyres vary enormously, depending on how they will be used. For example, a tyre intended for a wheelbarrow would be quite unsuitable for a Mini, even if the size was right. It is not just size that matters: the tyres must be right for the load carried by the vehicle, right for the speed at which it will move, and right for the different surfaces over which it will travel.

So how *do* you decide which tyres to fit on which vehicle? In practice this is fairly easy: each tyre manufacturer produces a list that tells you which tyres are suitable for which vehicle.

It may be easy to look up the tyres on a list, but you still need to know something about the various types of tyres. This knowledge will help you to answer customers' questions. They will see that you know what you are doing. In time, it will also make your job easier and more interesting.

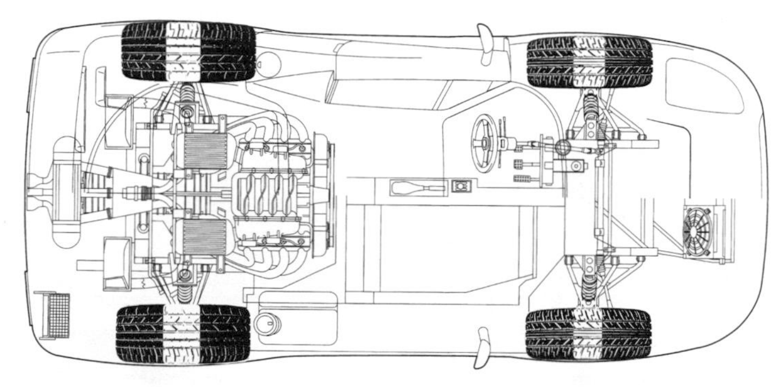

Your 500 hp Porsche may do 0–60 mph in only 4.0 seconds and reach 200 mph, but this is safe only if the tyres are the right sort and properly fitted. All that power and control depends on the four small areas where the rubber is in contact with the road.

QUESTION

List below *three* main factors to think about when you choose the tyres for a vehicle.

1 ....

2 ....

3 ....

ACTIVITY

Imagine that you are going to fit a new set of tyres to your favourite car. Complete the following table.

Car		**Tyre**			
Make	Model	Size	Make	Type	Cost

Tyre construction

Just glancing at the outside, all tyres look much the same. Under the outer layer of rubber, though, there are *two* main kinds of construction: **cross-ply** (or **diagonal-ply**) and **radial-ply**.

Cross-ply

This is the traditional form of tyre. The **casing** plies are made of textile cords: they run *diagonally* from **bead** to bead. The layers crisscross each other. These tyres tend to give a fairly stiff ride. Compared with radial tyres, this type is not as good at holding the road, and they may wear out more quickly. Generally they are made for older vehicles, and most are fitted with a tube.

cross plies
tread
apex strip
side wall
bead
inner lining
bead wires

1 ....

Radial-ply

In this tyre, the casing cords run *radially* from bead to bead. Notice that there is a 'belt' – maybe more than one – of textile or steel cords (or both) beneath the tread. This form of construction was introduced by Michelin in 1948. Since then it has become the choice for almost all cars and commercial vehicles. Because of the tread bracing it will go further before wearing out; it will use slightly less fuel; and (depending on the tread pattern) it will grip the road better.

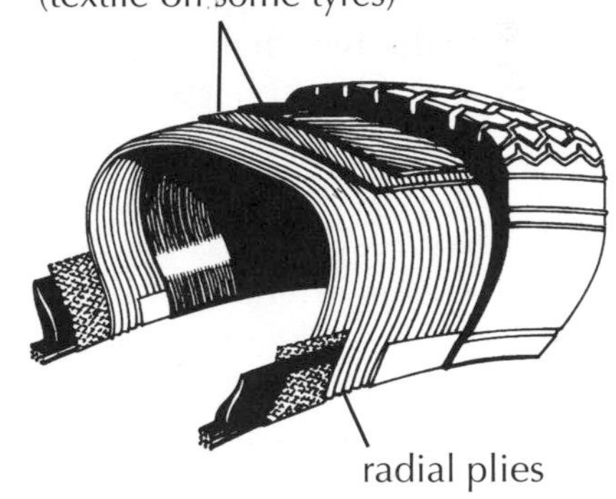

2 ....

QUESTION

Show the direction in which the plies would run in a radial-ply tyre.

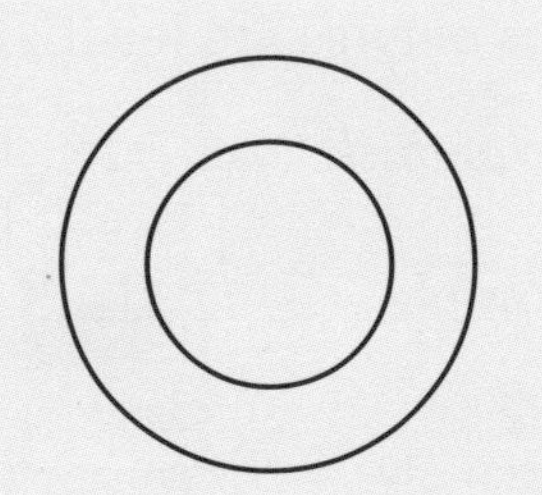

ACTIVITY

Look carefully at the two diagrams above. Look at how the **plies** are positioned. Then decide which is which. Check that you have got it right, then write in the correct titles.

Notice the names of the main parts of the tyres. You will need to remember these names.

Bias-belted

There is one other type of tyre construction, the **bias-belted tyre**. This is much less common. It is something of a mixture between cross-ply and radial-ply tyres. It has textile casing cords running diagonally, *and* a belt of textile cords under the tread.

Bias-belted tyres are not common in Britain. They are used on some farm machinery, on motorcycles, and on old American cars. Unless stated otherwise, this book will deal with cross-ply and radial-ply tyres.

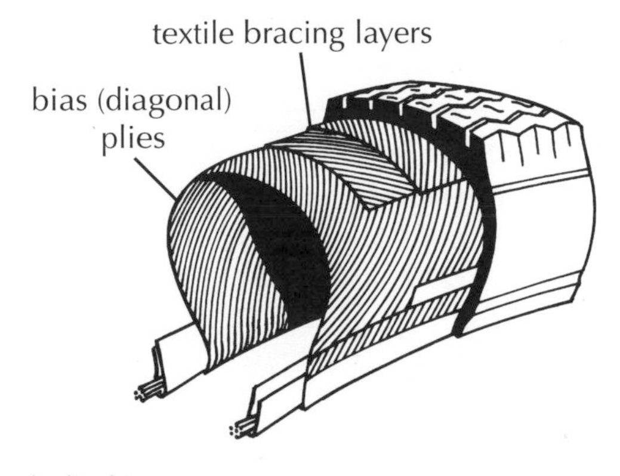

Bias-belted tyre

Tyres and the law

As a tyre technician, you need to become something of a legal expert! You must know and understand certain important parts of the law relating to tyres.

The only contact a car has with the road is through its tyres. Even if you drive a 500 bhp super-car, the tyres must grip properly. If a tyre punctures, or if it is the wrong sort of tyre for the driving conditions, then even the best driver may lose control. This could result in damage or even death.

It is therefore vital that every vehicle uses suitable tyres, and that these are properly maintained. To help make sure of this, there are legal guidelines: the **Motor Vehicle Construction and Use Regulations**. From time to time these are added to and updated.

Under these regulations **it is an offence, with severe legal penalties, for any person to use, or cause, or permit to be used on the road, any motor vehicle or trailer with defective or unsuitable tyres.**

There is no need to learn all of the legal points straight away. This book will explain the most important ones as each topic is dealt with.

TIP
Remember: these regulations affect *you*, as well as the job you do. If you ignore them, you might find yourself in serious trouble.

Mixing tyres

As you have seen, the two main types of tyre construction have different characteristics. As you can imagine, it can be dangerous to mix tyres on one vehicle. Some combinations are so dangerous that, in the UK, **most 'mixtures' of cross-ply and radial-ply tyres are illegal**. (Some drivers will not even mix similar tyres of different makes, because they might be different enough to affect the way the car handles.)

Radial and cross-ply tyres must never be fitted on the same axle. The *only* legal mixture of tyres on cars and light commercial vehicles is two cross-plies on the front, and two radials on the rear. (There is, however, a rare exception in the case of very wide tyres, where the tread is at least 300 mm wide on all the wheels.)

There is one *main* exception regarding mixing tyres. It *is* legal to fit a specially made **space saver** or **temporary spare**, just so that you can get home. With such a tyre fitted, you may not drive faster than 50 mph.

QUESTIONS

What are the *two* main types of tyre construction that you are likely to deal with?

1 ....

2 ....

Of these two, which type is fitted to all modern cars and light commercials?

....

ACTIVITY

Complete the drawings to show which types and positions of tyres are legal.

Car, van or truck

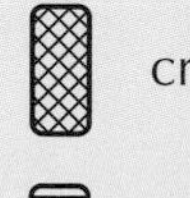

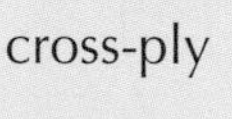

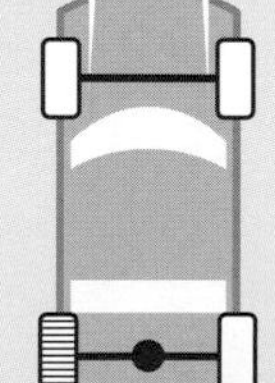

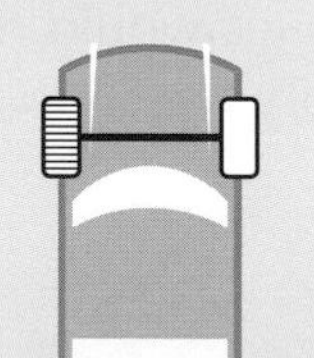

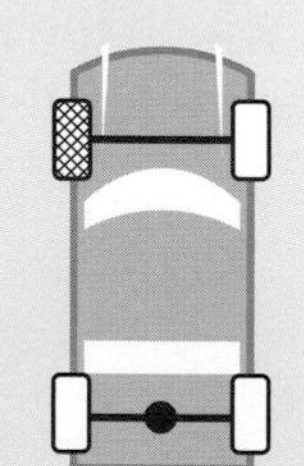

WORKSHOP

Look at a 'space saver' spare. List *three* ways in which it is clearly different from the other wheels and tyres on the car.

1 ....

2 ....

3 ....

The writing on the wall

When looking at a tyre, most people will notice the name of the manufacturer, because that is in big letters! A few will also notice the size. But other markings on the side tell the tyre technician a great deal about the tyre.

Look at the drawing below. The amount of information it contains seems enormous. Most of it is easy to follow, and the rest can soon be learned.

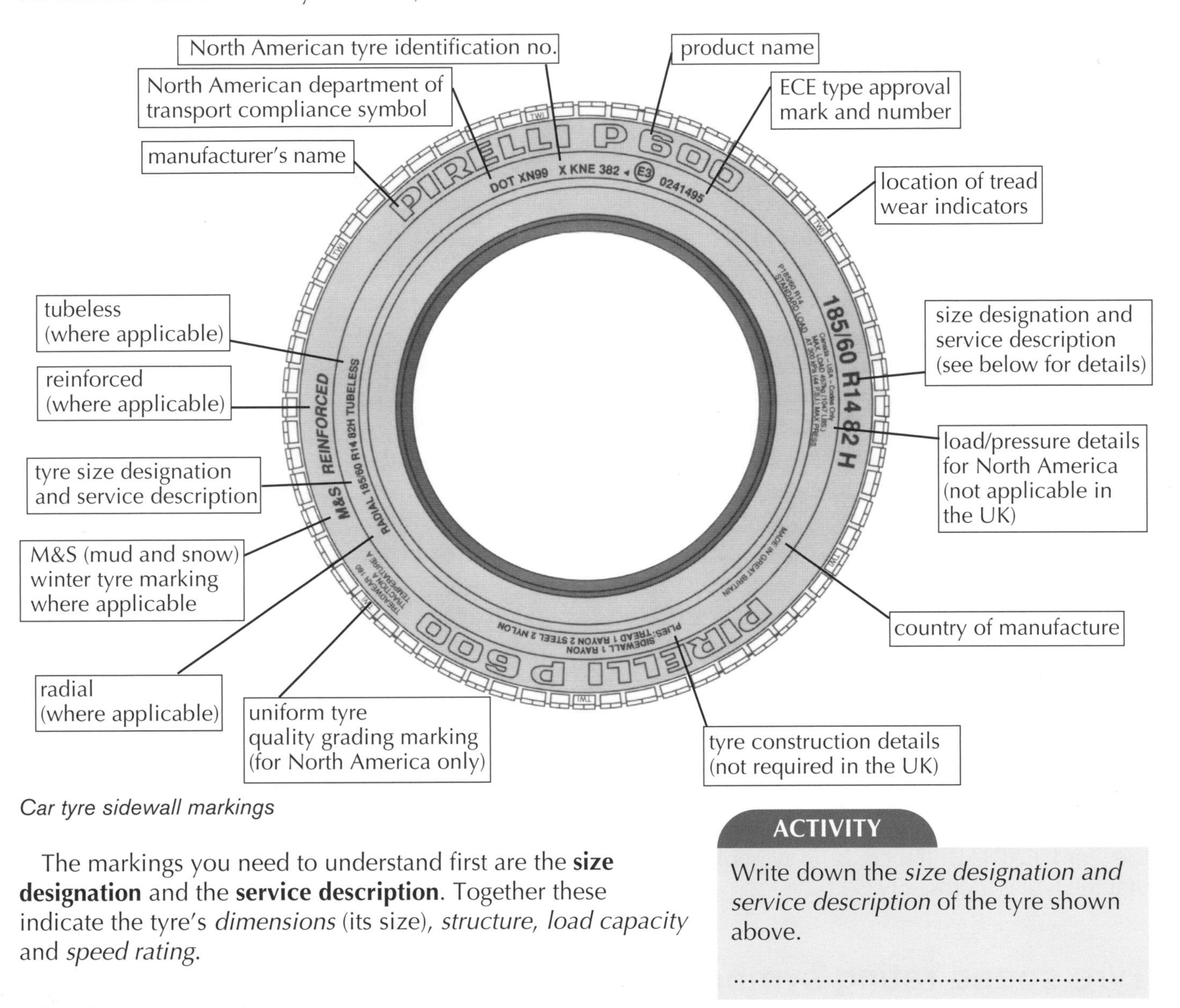

Car tyre sidewall markings

The markings you need to understand first are the **size designation** and the **service description**. Together these indicate the tyre's *dimensions* (its size), *structure, load capacity* and *speed rating.*

ACTIVITY

Write down the *size designation and service description* of the tyre shown above.

....

ACTIVITY

Look carefully at the drawing above. As you will see, some of the information does not apply in this country. Put a cross alongside each box that is not relevant in the UK.

The symbols in the drawing have these meanings:

Symbol	Meaning
185	Nominal section width (millimetres if a radial-ply tyre)
60	Tyre aspect ratio
R	Radial-ply construction
14	Nominal rim diameter (inches)
82	Load index (LI)
H	Speed rating symbol

It is important to look in rather more detail at what these markings mean.

Tyre section width and rim diameter

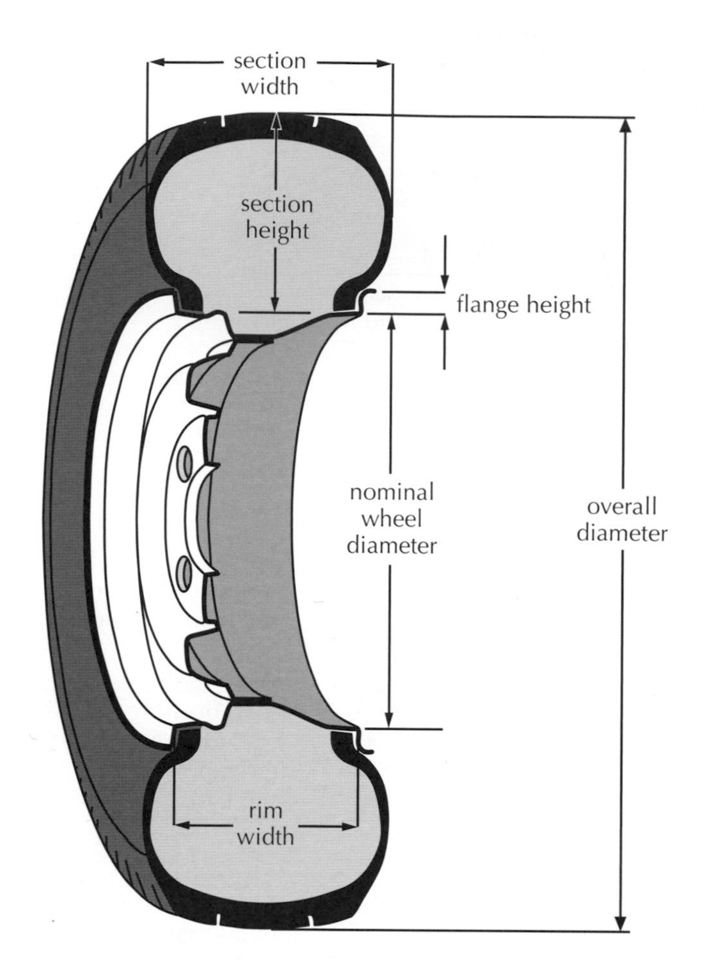

Tyre aspect ratio

This ratio shows the relationship between the section *height* and the section *width*. The ratio is expressed as a percentage. The normal or **standard aspect ratio** for radial-ply tyres is 82%.

Here are some examples.

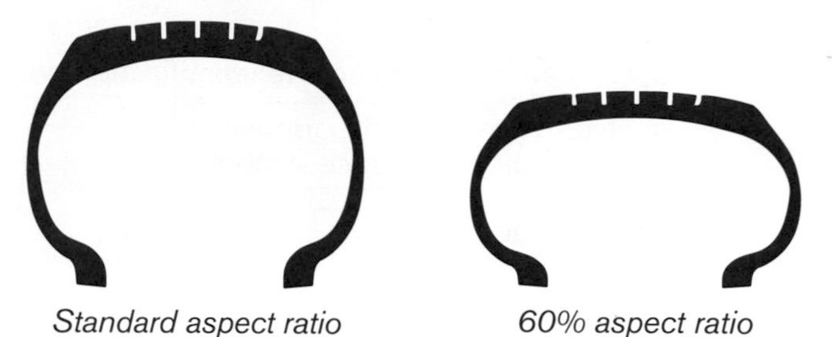

Standard aspect ratio

60% aspect ratio

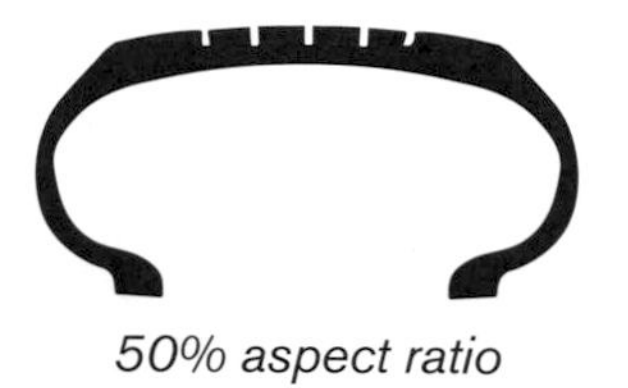

50% aspect ratio

Tyre construction

This letter tells you whether the tyre is of cross-ply or radial-ply construction.

Most modern tyres are radials. They carry the letter 'R' or the word 'RADIAL'. If there is *no* symbol, you may assume that the tyre is a cross-ply.

TIP

Radial-ply tyre: the section width is in millimetres.

Cross-ply tyre: the section width is in inches.

Tyre speed rating symbol

This letter tells you the maximum speed for any particular tyre. Usually it is shown at the end of the size designation.

Part of a speed symbol table is shown here. These are the most common symbols. Don't worry if you don't remember them right away – every tyre firm has a table so you will be able to check.

Q	For cars with max. speeds up to 160 km/h (100 mph)
R	For cars with max. speeds up to 170 km/h (105 mph)
S	For cars with max. speeds up to 180 km/h (113 mph)
T	For cars with max. speeds up to 190 km/h (118 mph)
H	For cars with max. speeds up to 210 km/h (130 mph)
V	For cars with max. speeds up to 240 km/h (150 mph)
W	For cars with max. speeds up to 270 km/h (168 mph)

ACTIVITY

Try reading the table. What is the speed rating for the 185/60 R 14 82H tyre shown on page 59?

It would be suitable for a maximum speed of:

.... mph.

HEALTH & SAFETY

Take care! Sometimes the speed symbol is near the *middle* of the service designation:

205/55 VR 15 or 225/45 ZR 16

Make sure you do not confuse it with the speed symbol 'V' at the *end* of the service designation:

195/65 R15 91V

The 'VR' and 'ZR' speed symbols are for tyres designed for very high-speed cars. Such tyres may be capable of over 150 mph, or even over 200 mph. There are special load and speed considerations for *all* 'V' and 'Z' rated tyres, and *no load index* is shown on the tyre.

V	For cars with max. speeds over 210 km/h (130 mph)
Z	For cars with max. speeds over 240 km/h (150 mph)

TIP

Play safe! If you have any doubts about symbols, check with your supervisor.

Tyre load index

This number states the *maximum* load that *one tyre* can safely carry. You can check this on a **load index (LI) table**. You do not need to remember this – there will be one in every tyre shop.

Table of Load Indices (L.I.) and Associated Loads

Load Index	Max. Load kg	Load Index	Max. Load kg	Load Index	Max. Load kg
60	250	84	500	108	1000
61	257	85	515	109	1030
62	265	86	530	110	1060
63	272	87	545	111	1090
64	280	88	560	112	1120
65	290	89	580	113	1150
66	300	90	600	114	1180
67	307	91	615	115	1215
68	315	92	630	116	1250
69	325	93	650	117	1285
70	335	94	670	118	1320
71	345	95	690	119	1360
72	355	96	710	120	1400
73	365	97	730	121	1450
74	375	98	750	122	1500
75	387	99	775	123	1550
76	400	100	800	124	1600
77	412	101	825	125	1650
78	425	102	850	126	1700
79	437	103	875	127	1750
80	450	104	900	128	1800
81	462	105	925	129	1850
82	475	106	950	130	1900
83	487	107	975	131	1950
				132	2000

ACTIVITY

What load would the 185/60 R 14 82H tyre carry?

....

ACTIVITY

Consider a tyre marked 155/80 R 13 79T. Fill in the following table, saying what each symbol means.

Symbol	Meaning
155	
80	
R	
13	
79	
T	

Tubed and tubeless tyres

Originally, all tyres had an **inner tube**. This helped to stop the air from escaping. Later, though, one-piece welded wheel rims were introduced, instead of spoked wheels. Tyre technology improved, too, and it became possible to do without the tube. **Tubeless tyres** are cheaper and lighter. More importantly, tubeless tyres lose pressure much more slowly.

Tubeless tyres are always marked 'TUBELESS' or 'TL'. Tyres without such markings need to be fitted with tubes.

TIP
When you fit a tubeless tyre, make sure the tyre bead and the wheel rim are *clean* and *undamaged*. (Dirt or damage may cause air leaks.)

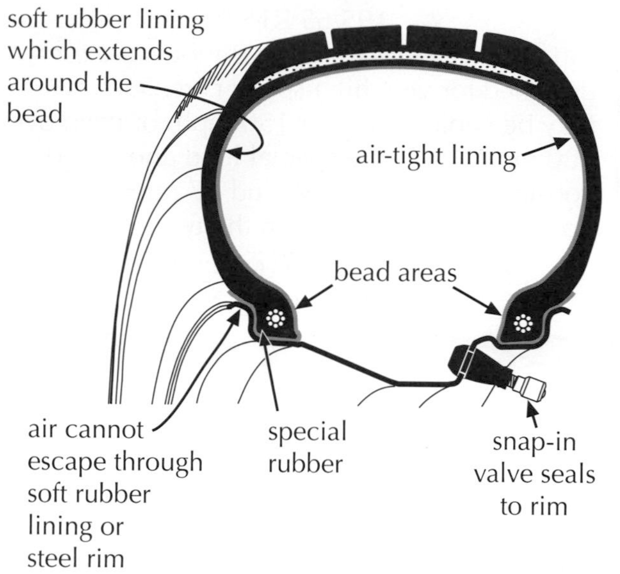

Tubeless tyre

Tubed tyres

Spoked wheels, some early car wheels, and some commercial vehicle wheels must be fitted with inner tubes. If in doubt, always check.

Occasionally inner tubes are fitted to tubeless tyres to stop a persistent, slow air loss. **Inner tubes must *not* be fitted to tyres of 65% (or less) aspect ratio, nor must they be used as a puncture repair.**

Leaks

In a tubeless tyre, the unstretched inner lining of the tyre would cling to the nail. This would make it hard for the air to escape. For this reason you should always inspect a tubeless tyre with great care – there may be something (such as a nail) which has damaged the casing, even though there is little or no loss of air.

ACTIVITY

If a tubed and a tubeless tyre each had a simple nail puncture, which would lose air quicker?

☐ tubed ☐ tubeless ☐ both equal

Tyre valves

A tyre valve does *three* jobs:

- it retains air (up to 250 psi)
- it allows air in, during inflation
- it allows air out, during deflation and pressure testing.

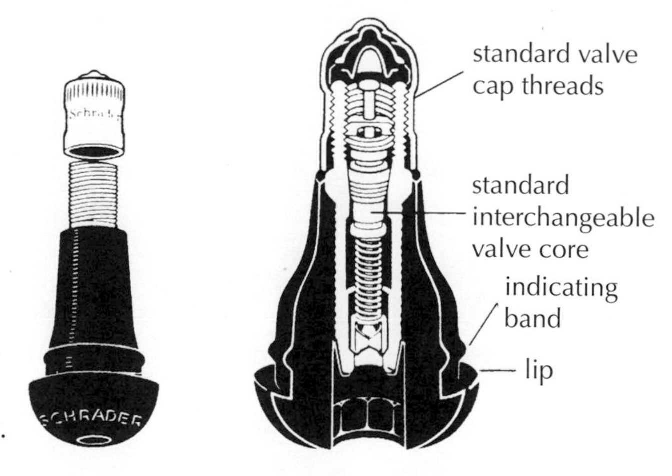

The type of valve shown on the previous page is the kind fitted with tubeless tyres for cars. Other valves may be metal-cased, as shown here, and may have a different fastening in the wheel, but the valve core will be just the same.

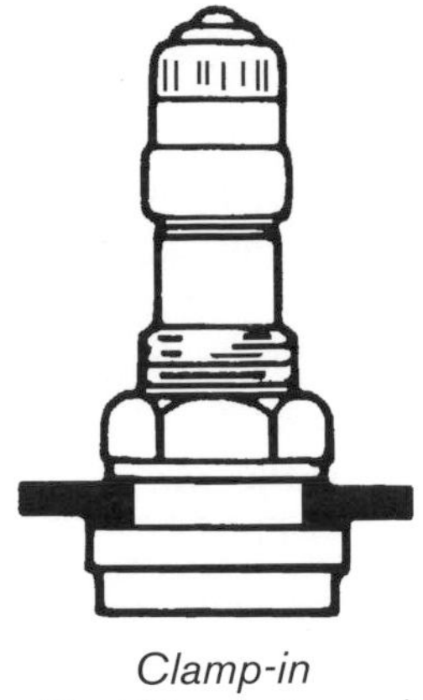

Clamp-in
(for V/Z rated tyres)

TIP

- Renew the **valve assembly** every time the tyre is removed.
- Metal **valve caps** retain pressure *and* keep out dirt. Plastic caps only keep out dirt.

Tyre pressures

The most common problem with tyres is that they are run at the **wrong pressures** – generally too low. What is more, **it is illegal to use a vehicle with wrong tyre pressures**. Be sure always to check the pressures!

Pressures that are *too high* can cause:

- excess wear at the centre of the tread
- a hard ride
- reduced resistance to impact damage
- a greater chance of skidding
- light steering.

Pressures that are *too low* can cause:

- excess wear on both edges of the tread pattern
- excess flexing and damage to the side walls
- 'sloppy' steering
- heavy steering
- poor road-holding.

QUESTIONS

1 Pressures should be checked when the tyres are:
☐ cold ☐ hot ☐ any temperature.

2 Suppose the pressures are set when the tyres are cold. If they are checked when the tyres are hot, they will be:
☐ higher ☐ lower ☐ the same.

3 If a vehicle's tyres are hot and the pressures are much higher than those recommended, should you:
☐ reduce the pressures?
☐ take no action?
☐ check when the tyres are cold?

4 List *two* places where you can find recommended tyre pressures.

a ....

b ....

5 List *two* effects that may be felt by the driver:

a when tyre pressures are too low

i ....

ii ....

b when tyre pressures are too high

i ....

ii ....

ACTIVITY

With many vehicles the recommended pressures may be different for front and rear tyres. They may also differ according to load and speed.

Fill in this table for *two* very different vehicles that you are allowed to check.

Vehicle: Make .... **Model**				
Tyre position	**Recommended pressure**			**Actual pressure**
	Light load	Full load	Max speed Full load	
N/S/F				
O/S/F				
N/S/R				
O/S/R				
Spare				

Vehicle: Make .... **Model**				
Tyre position	**Recommended pressure**			**Actual pressure**
	Light load	Full load	Max speed Full load	
N/S/F				
O/S/F				
N/S/R				
O/S/R				
Spare				

Which way round?

Directional tyres

The vast majority of tyres are **non-directional** – they may be fitted on any wheel of the car, and it does not matter which side of the tyre goes where.

Some specialist tyres, however, must be fitted so that the tread pattern always faces in a particular direction. Examples can be found on:

- very high-performance cars
- agricultural tractors
- earth-moving vehicles.

When fitting directional tyres, look on the sidewall for:

⇨ DIRECTIONAL

The arrow must point in the direction of *forward* rotation.

ACTIVITY

Look carefully at the two tyres shown. Which has the non-directional, and which the directional tread pattern?

....

Asymmetric tyres

An **asymmetric tyre** has a different pattern on each side of the tread's centre line. The two different halves of the tread have different road-holding qualities. For example, one half may give excellent wet grip, the other half may give excellent cornering and traction grip.

With some asymmetric tyres the tread may even be divided into *three* parts, as shown.

Such tyres are fitted only to very high-performance cars. **Asymmetric tyres must be fitted the right way round.** Look for markings on the sidewall such as:

SIDE FACING OUTWARDS SIDE FACING INWARDS

or

OUTER INNER

CENTRE gives maximum water clearance

INNER EDGE puts power onto road to give maximum traction

OUTER EDGE combats excessive wear from hard cornering and power steering

An asymmetric tyre (three parts)

Tyre faults

Tread faults

One of the jobs you will often have to do is to check tyres for faults.

QUESTION

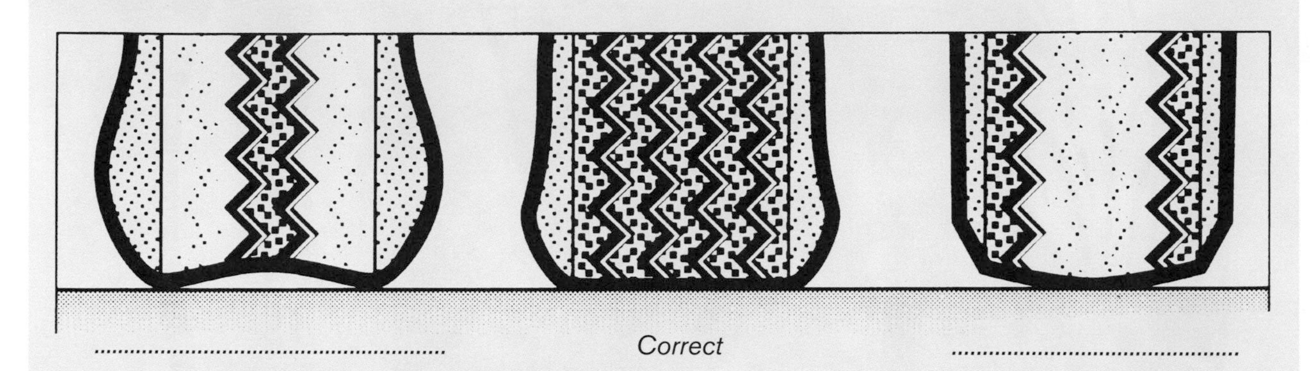

Which tyre is showing wear caused by under-inflation, and which by over-inflation?

Even if the tyre pressures are correct, there may be many other causes of uneven tyre wear. A good tyre technician soon learns to recognise them. Here are some common examples.

Edge wear

Too much **wheel camber** (tilt) can cause severe wear on one edge.

Flat spots

These can be caused by **brake grabbing** or **locking on**.

Feathering

When the edges of each tread groove are worn this causes a thin **feather** of rubber along the length of the groove. The most likely cause is the wheels being **out of alignment** (poor tracking).

Tread cut

The **cut** on the tread shown opposite has been there for a long time. It has let in water, rotting the plies and causing a piece of the tread to peel off.

Incorrect camber

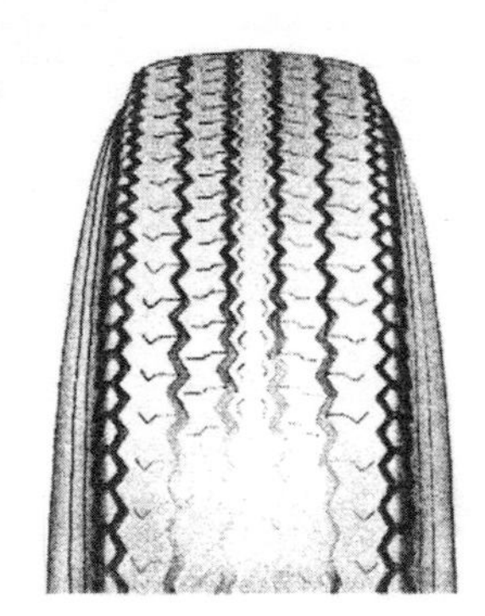

Flat spots

Feathering

Neglected cut

Wall faults

Bulges, lumps or tears

These are very likely to make the tyre useless. Often they are caused by hitting the kerb with the side of the tyre.

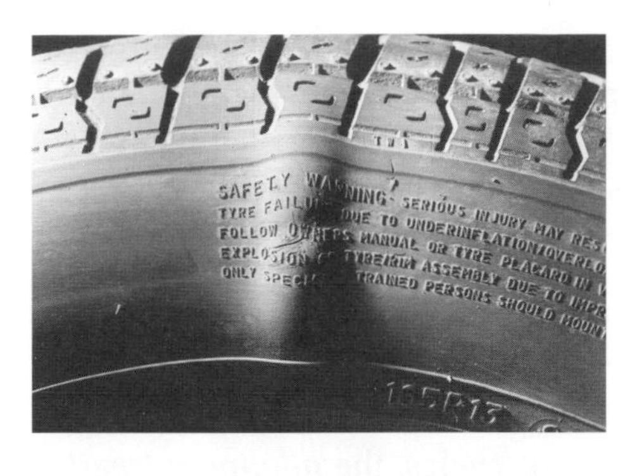

Bulge, due to kerb or pothole damage

Sidewall fracture or cut

This is usually caused by hitting a kerb or pothole very hard.

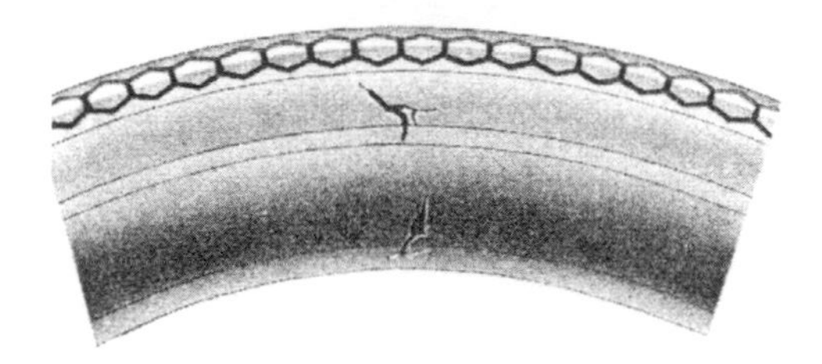

Sidewall fracture

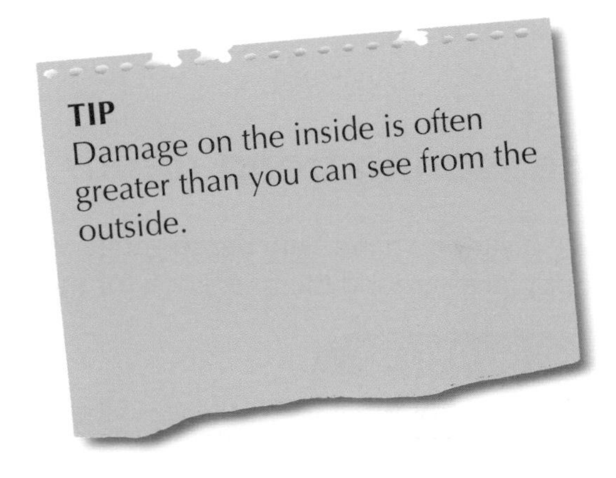

TIP
Damage on the inside is often greater than you can see from the outside.

The law

It is an offence to use a tyre that has

- **a cut over 25 mm (1 inch) long**, measured on the outside of the tyre, deep enough to reach the fabric
- **a bulge, lump or tear** caused by separation or failure of its structure.

Tread depth

Another job you will often be asked to do is to check the **tread wear**.

In the United Kingdom (and the European Union) there are very strict limits governing the *minimum depth* of tread on tyres. If you drive a vehicle with even one tyre too worn, you could be in serious trouble. For each defective tyre, you would be likely to get a heavy fine and licence endorsement.

One main purpose of the tread pattern is to give good grip in wet weather. The tread grooves channel away the water. This allows the tread blocks to grip the road, instead of sliding over a film of water. If the tread is worn, the water won't flow away. **Worn tyres can kill.**

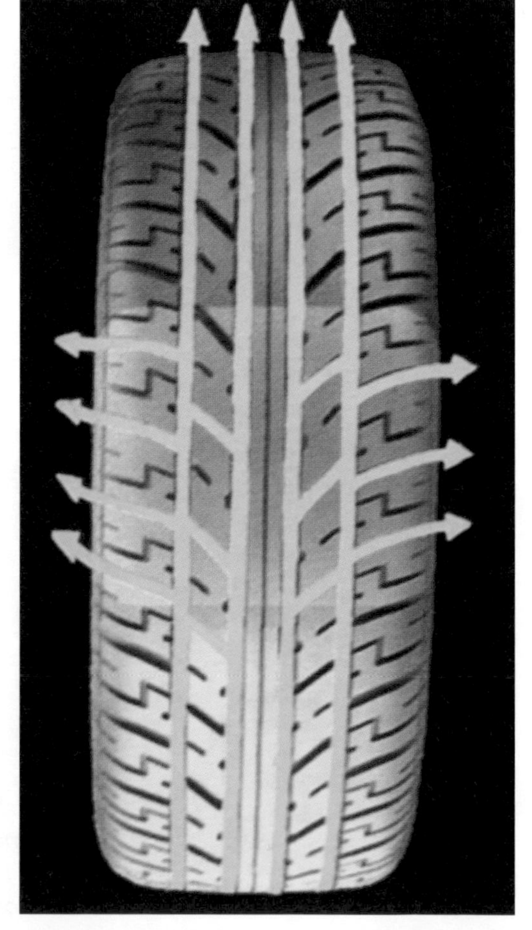

In wet weather water is channelled away from the contact patch, outward along the tread grooves. Some of it goes out at the sides

The law

The tread pattern of every tyre fitted to a car, light commercial vehicle (up to 3500 kg) or light trailer, shall be *at least 1.6 mm deep* throughout a continuous band comprising the central three-quarters of the breadth of the tyre and round the entire circumference.

Note: for goods vehicles over 3500 kg and motor cycles, the *minimum tread depth* must be 1.0 mm over three-quarters of the breadth of the tread, but with visible tread on the remaining quarter, and around the entire circumference.

The tread grooves must go all the way around (circumference) in a continuous (unbroken) band

Measuring tread depth

Use a simple, hand-held **depth gauge**. Always measure at least four places around the tyre. Look for the *minimum* readings.

ACTIVITY

Find a set of worn tyres. Measure the treads, and fill in the following table.

Vehicle: Make .... Model ....

Tyre position	Make and size	Tread depth	Condition
N/S/F			
O/S/F			
N/S/R			
O/S/R			
Spare			

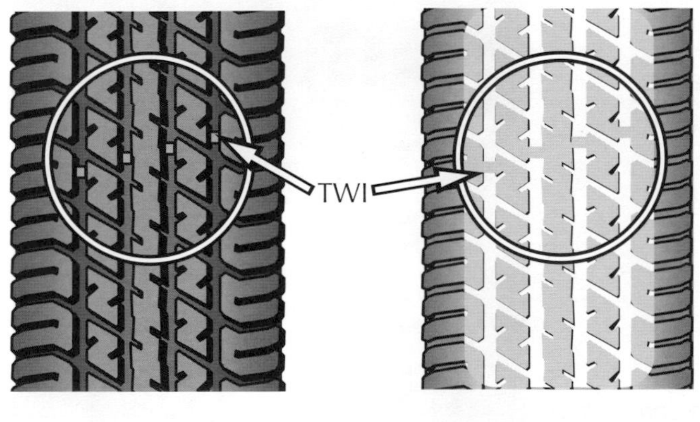

New tyre

Worn tyre: This tyre is very close to being – or has just become – illegal

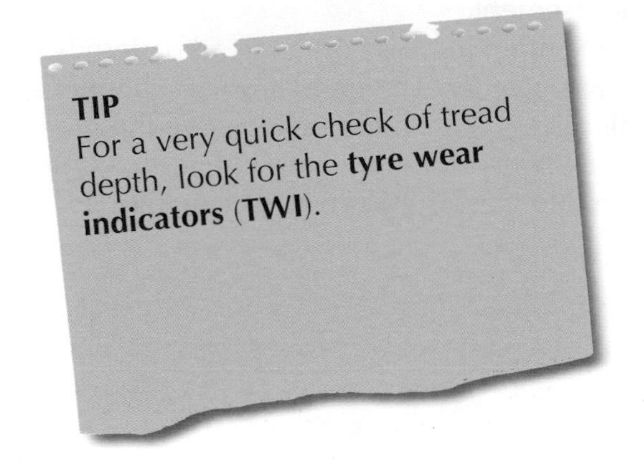

Wheel balance

Wheel out of balance

When customers complain of **vibration** felt through the steering wheel, usually at about 45–50 mph, the reason is probably that the wheel is **out of balance**. If it is really bad, the whole car may shake.

A skilled tyre technician can often spot the trouble simply by looking at the tyre tread. A wheel that has been out of balance for a long time will often show uneven patches of wear.

Static balance

ACTIVITY

You probably already know more about wheel balance than you realise. For example, suppose that you lifted the front of a bicycle and allowed the wheel to spin freely. Where would the valve be when the wheel stopped? Mark it on the sketch.

The wheel would stop with the valve at the bottom. The valve adds to the overall weight, so this is the heaviest part of the tyre/wheel assembly. The assembly is out of balance. If you rotated the wheel quickly, the whole bicycle would bounce up and down! You can imagine how much worse this would be with a car wheel at speed.

If you were to wind some lead wire (about the same weight as the valve) around the spokes *opposite* the valve, the spun wheel would not stop at any particular point. The assembly would be in **static balance**. At speed, it would spin smoothly.

This is roughly what is done with motor vehicle tyre and wheel assemblies. In this case, though, solid weights are clipped or stuck to the wheel rim to give balance.

Dynamic balance

Motor vehicle tyres are much wider than those on bicycles, however. As a result, static balance on its own will not cure *all* wheel balance problems. The wheels also need **dynamic balance**.

Two equal heavy weights placed as shown below could give static balance, but not dynamic balance. As it rotated, the wheel would tend to rock on the steering pivot.

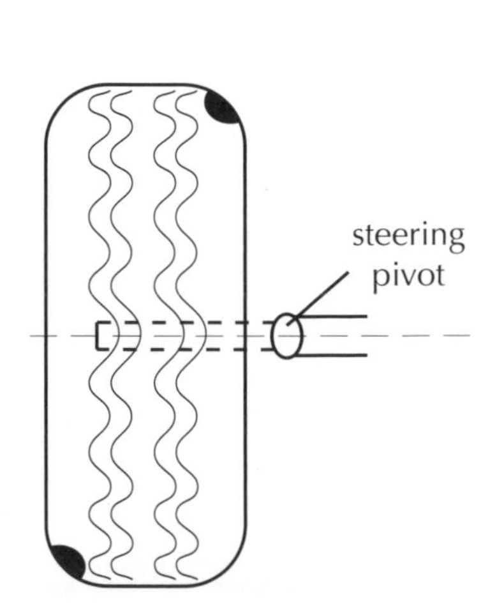

ACTIVITY

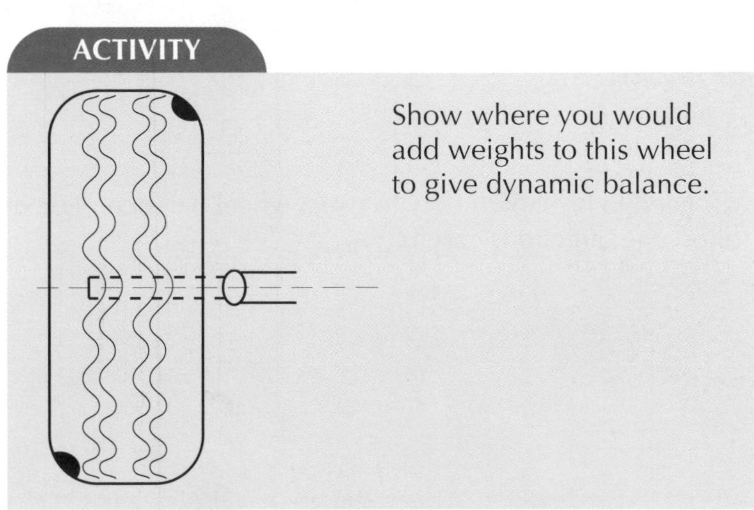

Show where you would add weights to this wheel to give dynamic balance.

Tyre/wheel balancing equipment

There are many types and makes of **wheel balancing machines**. Each will:

- spin the wheel (sometimes quite quickly)
- show what weights are needed
- show where they need to be fitted.

The machines are of two kinds:

- for balancing *off* the vehicle
- for balancing *on* the vehicle.

The second type also balances the rotating hub and brake drum or disc. However, it is not always possible to balance the wheel on the vehicle.

QUESTION

When is it necessary to balance tyre and wheel assemblies?

- ☐ Following a puncture repair.
- ☐ After fitting a new inner tube.
- ☐ When fitting a brand new tyre to a brand new rim.
- ☐ Following any removal of the tyre from the rim.
- ☐ Only when the customer complains of steering vibration.

On-vehicle balancer

Off-vehicle balancer

You will need to be shown how to use a wheel balancer. For your own safety, follow the directions carefully.

HEALTH & SAFETY

When the wheel spins, dirt or grit may fly out towards your face and eyes. Keep to one side to avoid it.

Tyre repairs

Most tyre depots repair punctures to **British Standard AU 159E**. This standard states very exactly the types and sizes of puncture that can be repaired, in which parts of the tyre, and *how* each is to be repaired – see the drawings on the right.

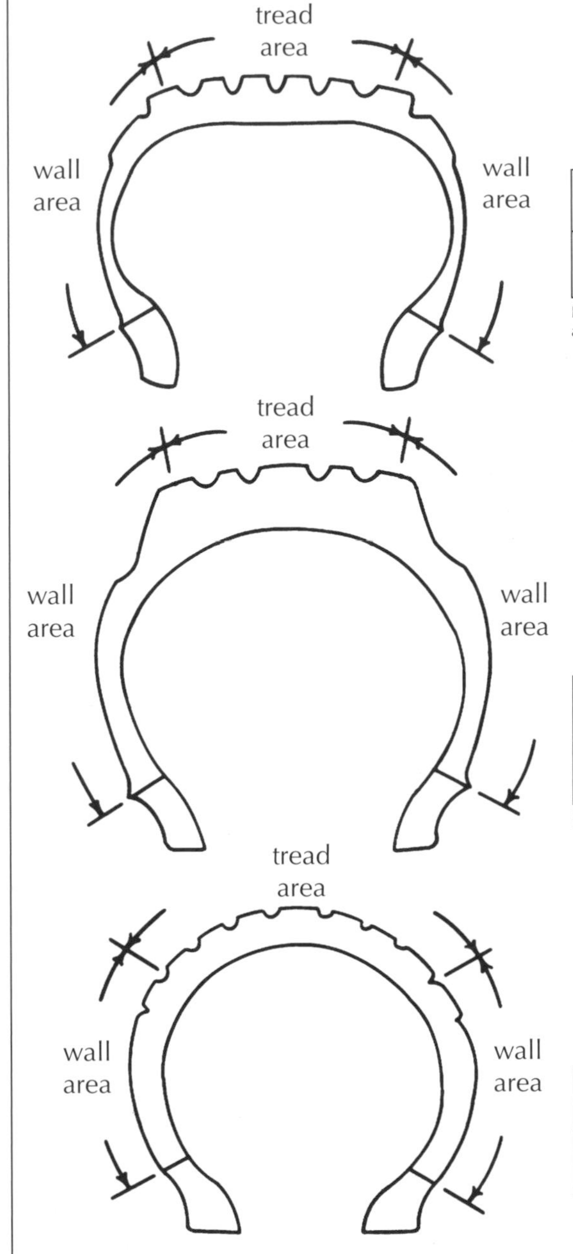

Car and light van

Cross-ply and radial-ply	Tread area	Wall area
All sizes and speed ratings	6 mm max	not allowed

Note: 'Z' rated tyres only 2 repairs per tyre advised. All others 3 repairs.

Commercial vehicles

Radial-ply	Tread area	Wall area
All sizes	10 mm max	not allowed

Note: only 4 repairs per radial tyre advised.

Motor cycle/scooter

Cross-ply and radial-ply	Tread area	Wall area
All sizes and speed ratings	6 mm max	not allowed

Note: only 2 repairs per tyre advised.

The tread area in a motor cycle/scooter is taken to be three-quarters of the nominal section width.
Note: no repairs are allowed in the bead area of any tyre.

Car tyres – repairable limits

As you can see, under British Standard AU 159E only fairly small punctures in the tread may be repaired. These repairs are all that most tyre depots would undertake.

Larger repairs can be made by specialist firms, but still within very strict limits. These firms must be approved by the **National Tyre Distributors Association** (**NTDA**). The drawing below, from the Goodyear *Radial Truck Service Manual*, gives some idea of what repairs can be done.

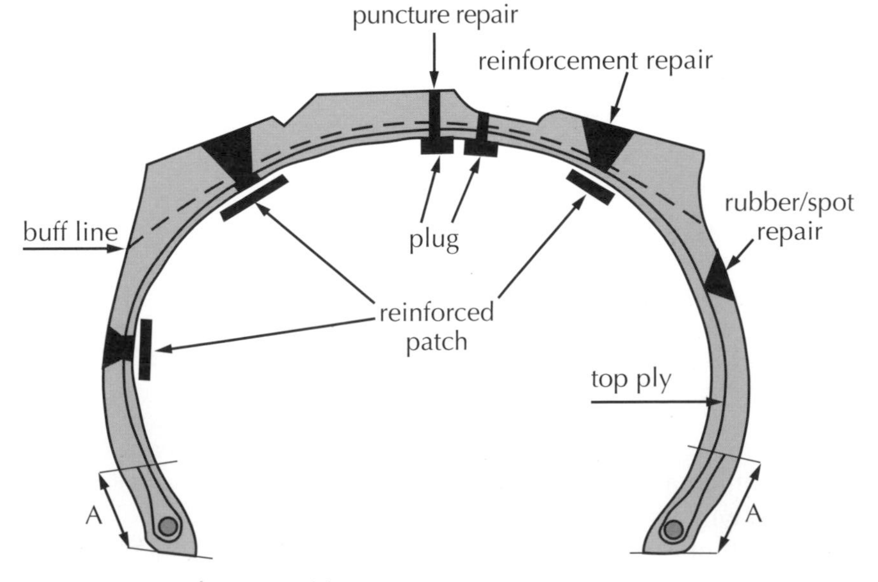

A = non-repairable area

Costs – repair or renew?

ACTIVITY

1 Consider if you had two tyres, each with a nail hole in the sidewall. One is from a small family car; the other is from a heavy commercial vehicle. With help from your supervisor, fill in this table.

Tyre	Size	Typical repair cost	Cost for new
Car			
Commercial			

2 If both tyres were half worn, what would be your advice to the customer?

Car tyre ☐ repair ☐ renew
Commercial tyre ☐ repair ☐ renew

Before deciding whether a tyre can be repaired, it *must* be taken off its rim so that the inside can be inspected. What from the outside looks like a simple nail puncture may have caused much greater damage inside.

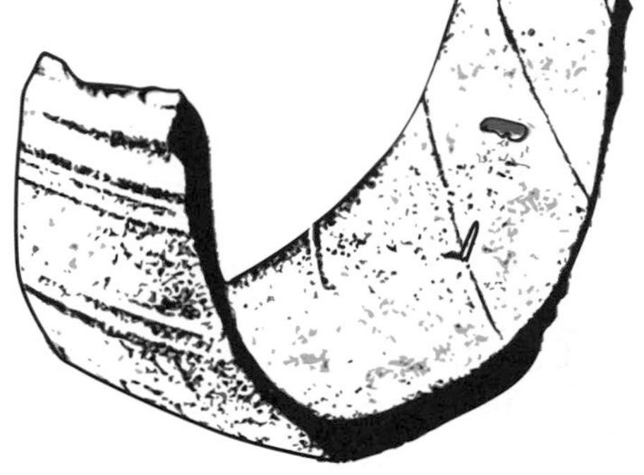

Secondary damage caused by a nail

QUESTION

Suggest *two* faults, not obvious from outside, that could seriously damage a tyre.

1 ....

2 ....

Repair methods

You will be shown in the workshop how to repair tyres. Every tyre repair that goes *through* the tyre must be **vulcanised** to the inside of the casing.

Vulcanising is done by heating the patch and the repair area of the tyre. This causes the patch and the tyre to 'weld' together. **Chemical cure** patches are an alternative, but the purpose is the same – to 'weld' the parts together.

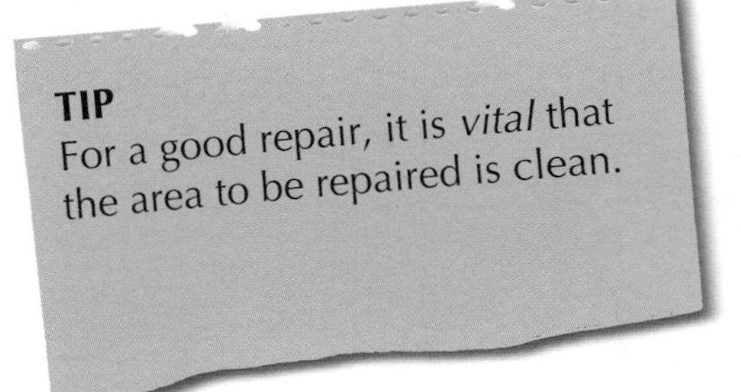

TIP
For a good repair, it is *vital* that the area to be repaired is clean.

Tube repairs

Many tyre firms recommend fitting a *new* tube, even if the puncture is small. If you *repair* tubes, you will again work to British Standard AU 159E. This standard sets limits to the amount of damage – a tube *more* damaged than this must be replaced.

Type of tube	Round hole: diameter	Patches	Split or tear	Patches
Motor cycle or scooter	10 mm	3 times diameter of damage	20 mm	$1\frac{1}{2}$ times length of split
Car, up to 7.00″	15 mm		40 mm	
Truck, above 7.00″	30 mm		60 mm	
Tractor	25 mm		100 mm	

As with tyre repairs, it is *vital* that the tube is clean and properly prepared.

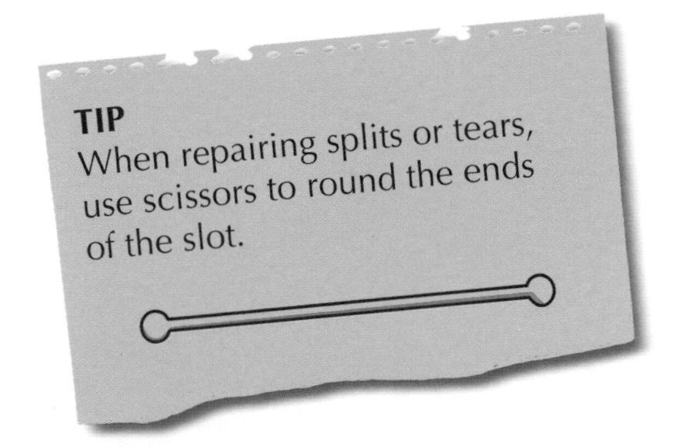

TIP
When repairing splits or tears, use scissors to round the ends of the slot.

QUESTION

When should wheel balancing be recommended to customers after a puncture repair?

- ☐ Every time, without fail.
- ☐ Not if the repair is small.
- ☐ Only if the workshop has little work.

WORKSHOP
Inspect some repaired tyres and tubes, to see what repairs can be done.

Light vehicle wheels

For cars and light commercial vehicles, there are three types of road wheel, as shown below.

QUESTION

Name these wheels, and say in each case what type of vehicle uses the wheel.

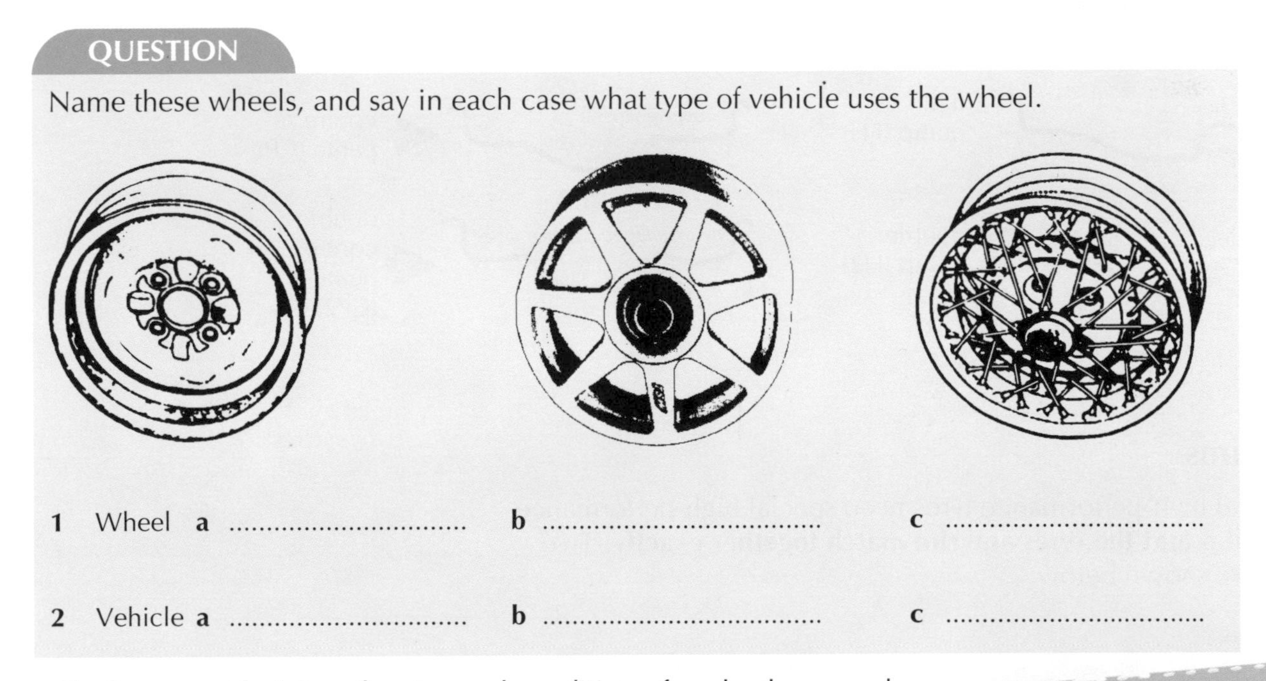

1 Wheel **a** .... **b** .... **c** ....

2 Vehicle **a** .... **b** .... **c** ....

To the tyre technician, the size and condition of a wheel are much more important than its looks. There are many differences between wheels, sometimes quite small differences. These can be very important. Always check the *rims*, which must be of the correct size and type. For example, if the correct size of tyre is fitted, but to the wrong type of rim, this could cause an accident.

TIP
Some modern wheels are *directional*, to help air flow to the brakes.

Well-base rims

A great variety of rims is available. A few examples only are shown here. At first it may seem that there is little difference between them, but you will soon learn to recognise them. As always, if you have *any* doubt, ask your supervisor.

The **well** allows one side of the tyre to drop into it during fitting or removal of the tyre.

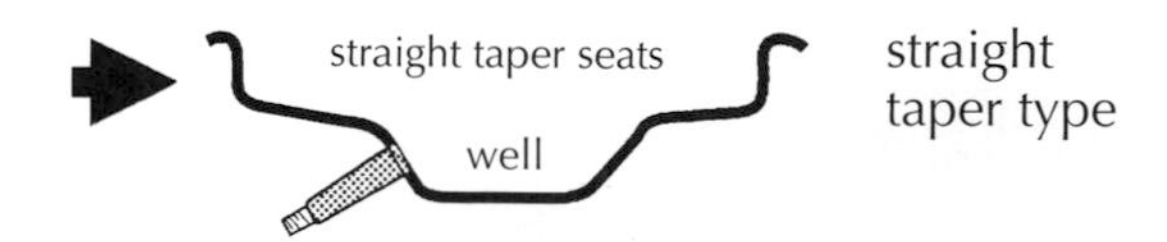

An old-type rim. This sort needs a tube, even if tyre is marked 'tubeless'

Modern rims

Below is a selection of modern rims, suitable for tubeless radial tyres.

Notice the **humps** close to the outer edges of the rims. These are safety features: they help to keep the tyre secure on the rim.

It is important to remove and to fit tyres from the correct side of the rim. The *arrows* show which side this is. **If you remove or fit the tyre from the *wrong* side, you might damage the tyre, and you might injure yourself.**

QUESTION

In what unit are these rim diameters measured?

....

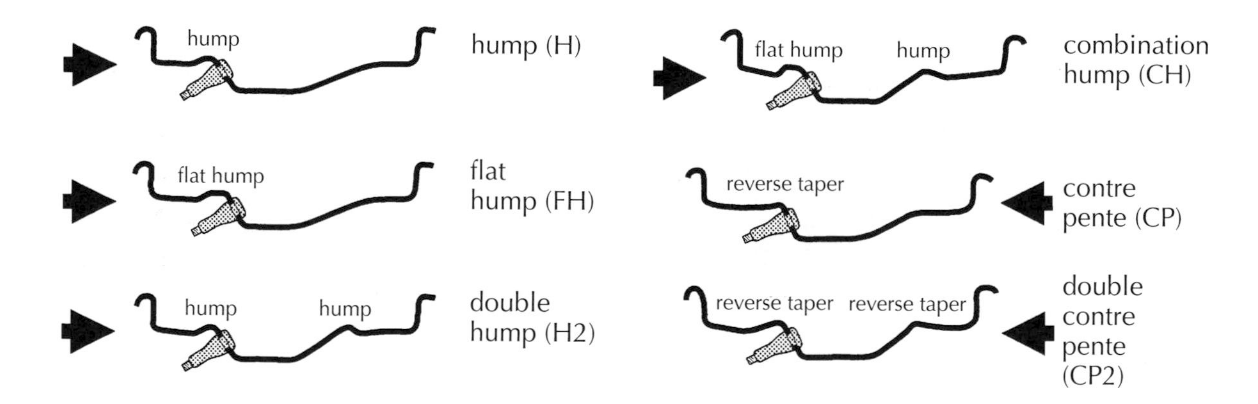

Special rims

Some special high-performance tyres need special high-performance rims. **It is vital that the tyres and rim match together exactly.** Two examples are shown below.

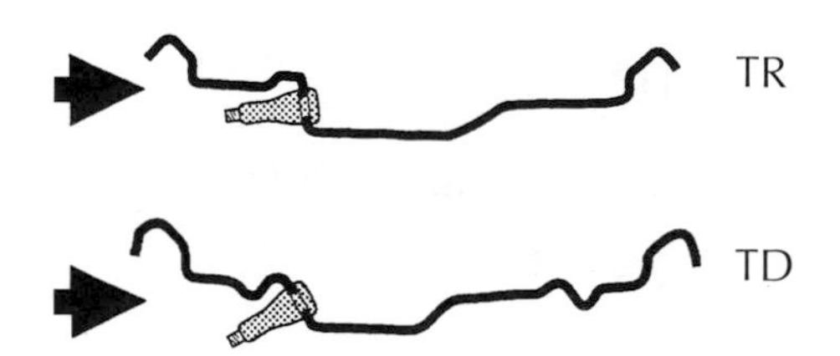

QUESTION

In what unit are these rim diameters measured?

....

Wheel rim measurement

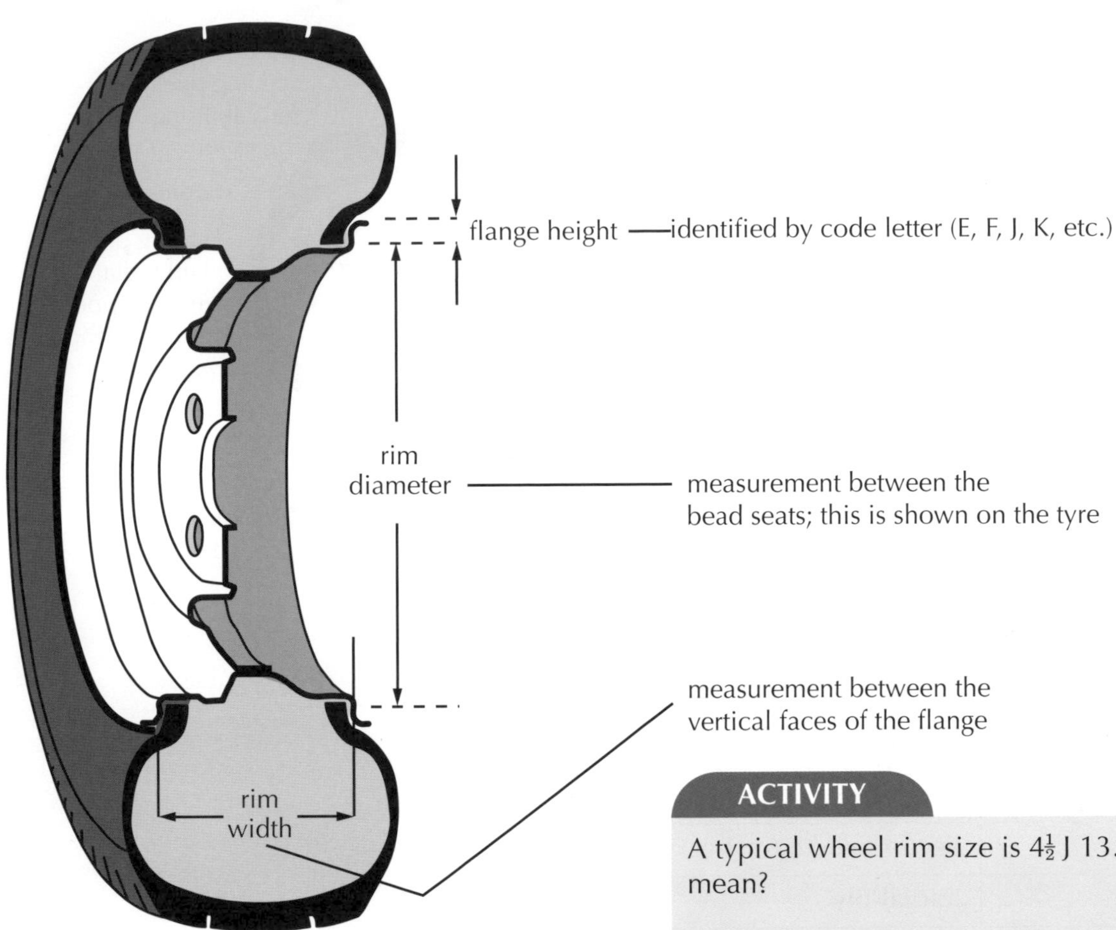

ACTIVITY

A typical wheel rim size is $4\frac{1}{2}$ J 13. What does that mean?

$4\frac{1}{2}$ ....

J ....

13 ....

Flange height code	Flange height	
	inches	mm
J	0.68	17.3
JK	0.71	18.0
K	0.77	19.6
L	0.85	21.6

Wheel condition

For safety, wheels must be checked. What condition are they in?

QUESTION

Apart from obvious impact damage, what *two* other things should you look for?

1 ....

2 ....

TIP
Sometimes fine cracks are very hard to see. Try holding the wheel clear of the ground and tapping it with a spanner. An uncracked wheel will give a bell-like ring.

Heavy commercial vehicle tyres

Even with commercial vehicles, tubeless radial-ply tyres are now the most popular type. They were introduced commercially by Michelin in the early 1960s.

They are heavier-duty tyres than those for cars, but they are of very similar construction and most are mounted on well-base rims.

tread rubber
bracing plies for stabilising the tread
shoulder
air-tight lining
casing plies laid radially and wrapped around beadcores
side wall
bead protector
reinforcing ply
rim
15° taper bead seat (to ensure that the rim is airtight)
special clamp-in valve
well-base rim

Size markings

In many ways size markings are very similar to those on car tyres, for example:

12 R 22.5
Tubeless 146/143.

The symbols have these meanings:

Symbol	Meaning
12	Section width (inches)
R	Radial-ply
22.5	Rim diameter (inches)
Tubeless	No tube needed
146	Load index – single tyre
143	Load index – twin tyres

TIP
Some older-size tyres use the ply rating (e.g. PR16) instead of the load index figure. Look at a ply rating table to find out what load the tyre can carry.

Two-piece and three-piece wheels: detachable-rim wheels

Some commercial vehicles do not use well-base rims. Instead, one of the rim flanges is **detachable**. This is removed to let the tyre and tube assembly slide off the wheel.

HEALTH & SAFETY

Before removing the rim or the locking ring, the tyre must be completely deflated. Remove the tyre valve and wait for all the air to escape.

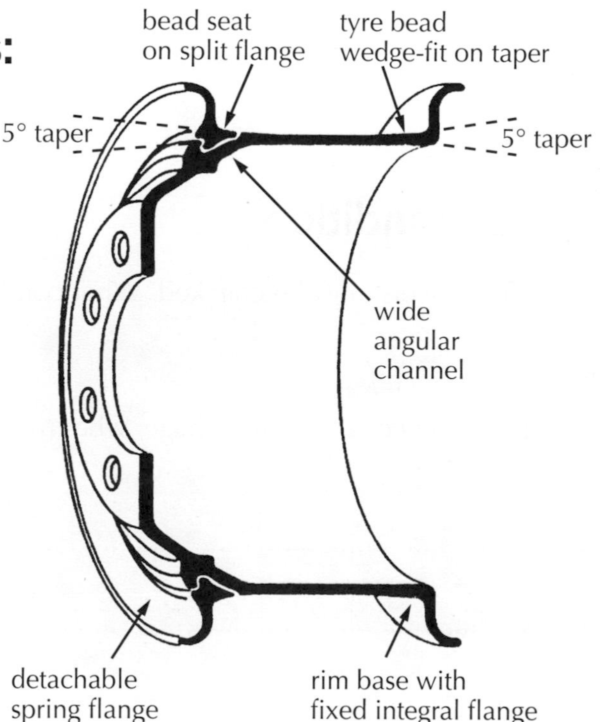

Two-piece rim

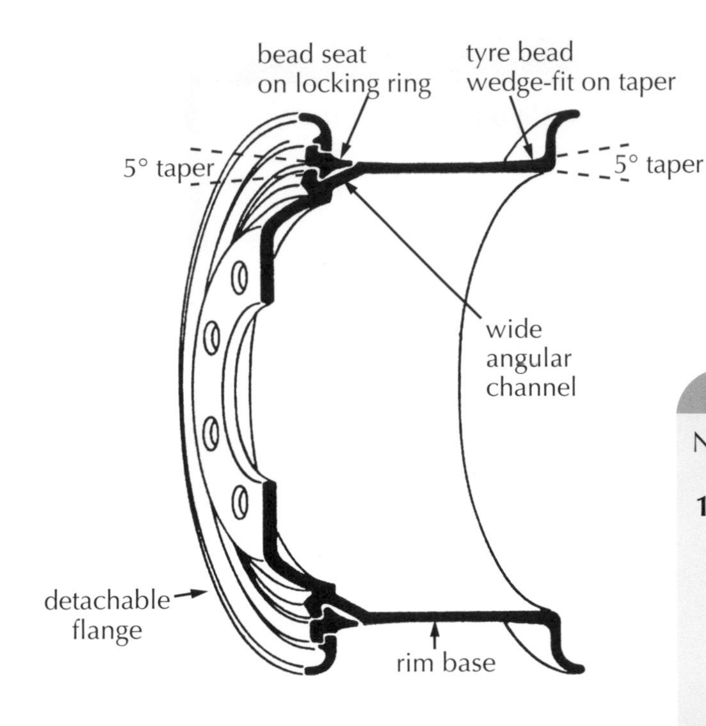

Three-piece rim

Typical rim markings

Symbol	Meaning
B	British Standard wheel
6.5	Rim width (Inches)
20	Rim diameter (inches)

QUESTION

Name the parts of this heavy-duty rim.

1 ....

2 ....

3 ....

Wheel offset

Look at the centres of several road wheels. You will see that the centre section, where they fasten to the hub (or brake drum), stands out on some and is set back (recessed) on others. This position is called the **offset**.

The offset is part of the vehicle design. It may give space for the brake drum or disc; it may affect the steering geometry; it may affect how the load is carried by the wheel. If the vehicle has twin (dual) tyres, the offset fixes the spacing between them.

QUESTION

Suggest *three* problems that might occur if the wheels fitted to a vehicle had the wrong offset.

1 ....

2 ....

3 ....

Measuring offset

As you can see from the drawings, the offset is the distance from the centre line of the wheel to the inside face of the wheel (where it comes up against the hub or drum).

- **Positive offset** – the inside face is *outside* the centre line.
- **Negative offset** – the inside face is *inside* the centre line.
- **Zero offset** – the inside face is *exactly on* the centre line.

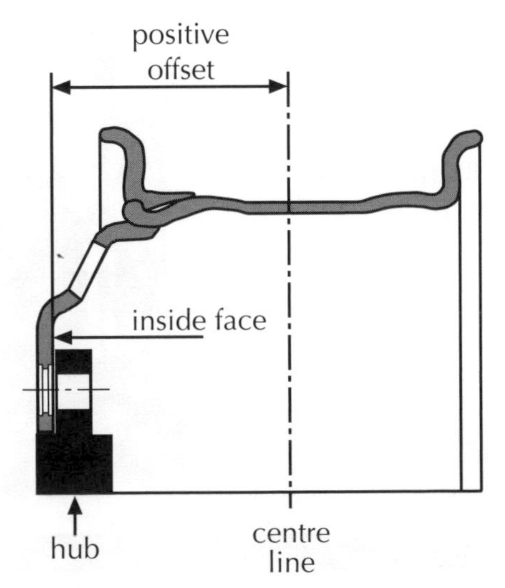

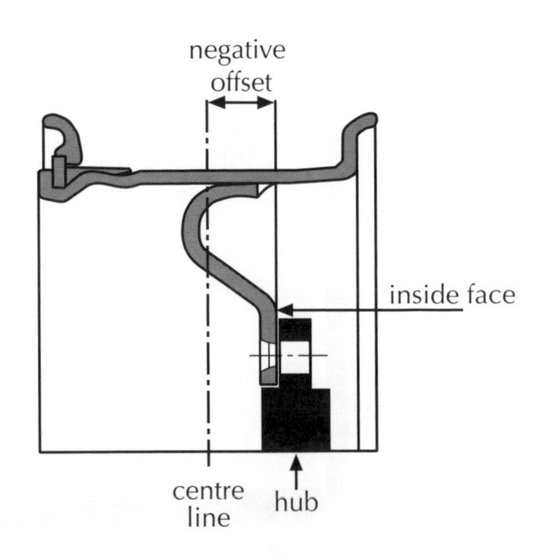

HEALTH & SAFETY

When fitting detachable rims, you must make certain they are *properly seated.*

1 **With the tyre in a safety cage**, inflate the tyre to about 15 psi (1 bar). Check the seating of the rim (and the locking ring, if any).
2 **With the tyre still in the safety cage, and while standing well to one side**, inflate the tyre to the correct pressure.
3 Check the rim seating again before removing the tyre from the cage.

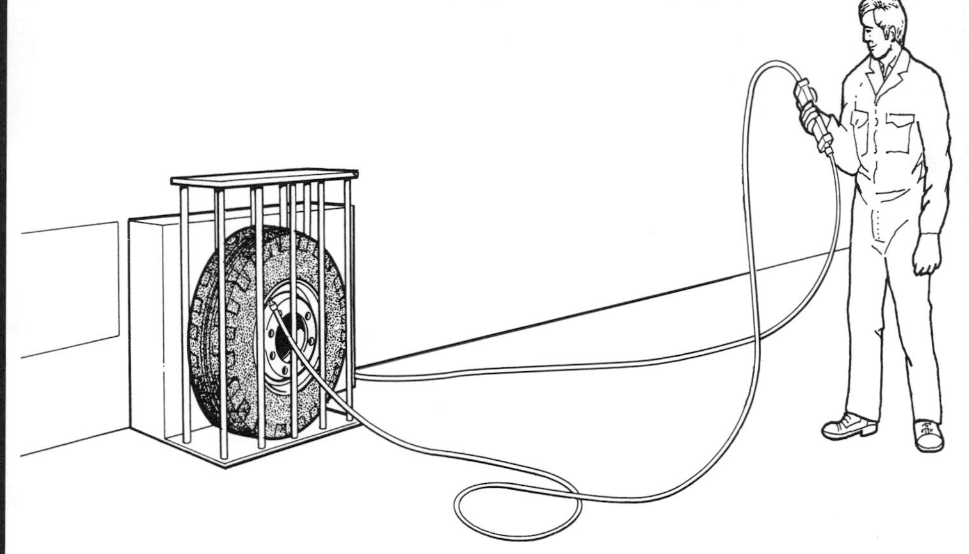

Inflating a commercial tyre using a cage

Beware! If a fully inflated large commercial tyre blows off its rim, it can explode with as much energy as a hand grenade. **Don't be tempted to find out!** You probably would not live to tell the tale.

HEALTH & SAFETY

Large tyre and wheel assemblies can be very heavy. You will be shown how to handle them without injury to yourself.

Lifting and transporting them can be made easy with this type of trolley.

Divided wheels

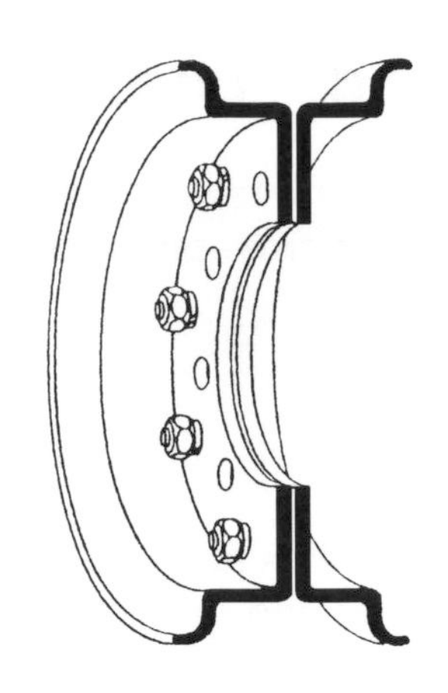

Divided wheels are easy to recognise: they have what look like *two* rings of wheel nuts. As you can see from the drawing opposite, it is the outer nuts that hold together the two halves of the wheel.

Removal

The tyre is removed and fitted by taking off the outer ring of nuts and splitting the wheel into its two halves. Do **take great care**: working on these wheels can be dangerous, especially if they are old.

HEALTH & SAFETY

Wheel already off vehicle

Before starting the job, remove the valve core and **make sure the tyre is completely deflated**.

HEALTH & SAFETY

Wheel still on vehicle

1. **Before removing the wheel** from the vehicle, remove the valve core and **wait until all the air has come out.**
2. **Only then** remove the wheel from the vehicle, by taking off the *inner* ring of nuts.

Replacement

After repairing the tyre or tube, or fitting a new tyre or tube, the two halves are bolted together again.

HEALTH & SAFETY

1. **Before inflation**, make sure all the *outer* (clamping) nuts are fully tightened.
2. **Place the assembly in a safety cage.**
3. Inflate to only 15 psi (1 bar). Check that the beads are seated.
4. Secure the assembly to the vehicle.
5. **Only then** inflate the tyre to the correct pressure.

QUESTION

List *three* types of vehicle that might use divided wheels:

1 ....

2 ....

3 ....

Flaps

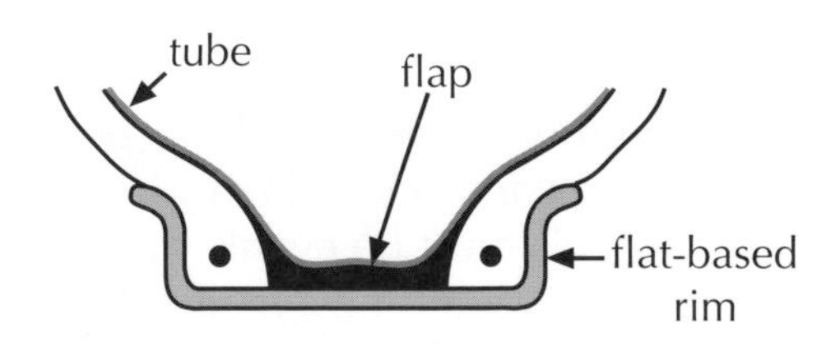

Flat-based rims normally have tyres fitted with an inner tube. A rubber **flap** is used to protect the tube from **chafing** on the rim. The flap is positioned as shown in the drawing opposite.

Note: one flange must be detachable.

Tyre mixing – commercial vehicles

As with cars, there are regulations about how tyres may be mixed on a commercial vehicle. **Radial and cross-ply tyres must never be mixed on the same axle.**

ACTIVITY

1 Look at wall charts and discuss tyre law with your supervisor.
2 Then complete the drawings below to show the permitted (legal) tyre combinations. Take care – this is not easy! If you get it wrong on a vehicle, driving it could be dangerous (and you could be in trouble).
Note the wide-tyre exception mentioned on page 10. (Many commercial tyre treads are over 300 mm wide.)

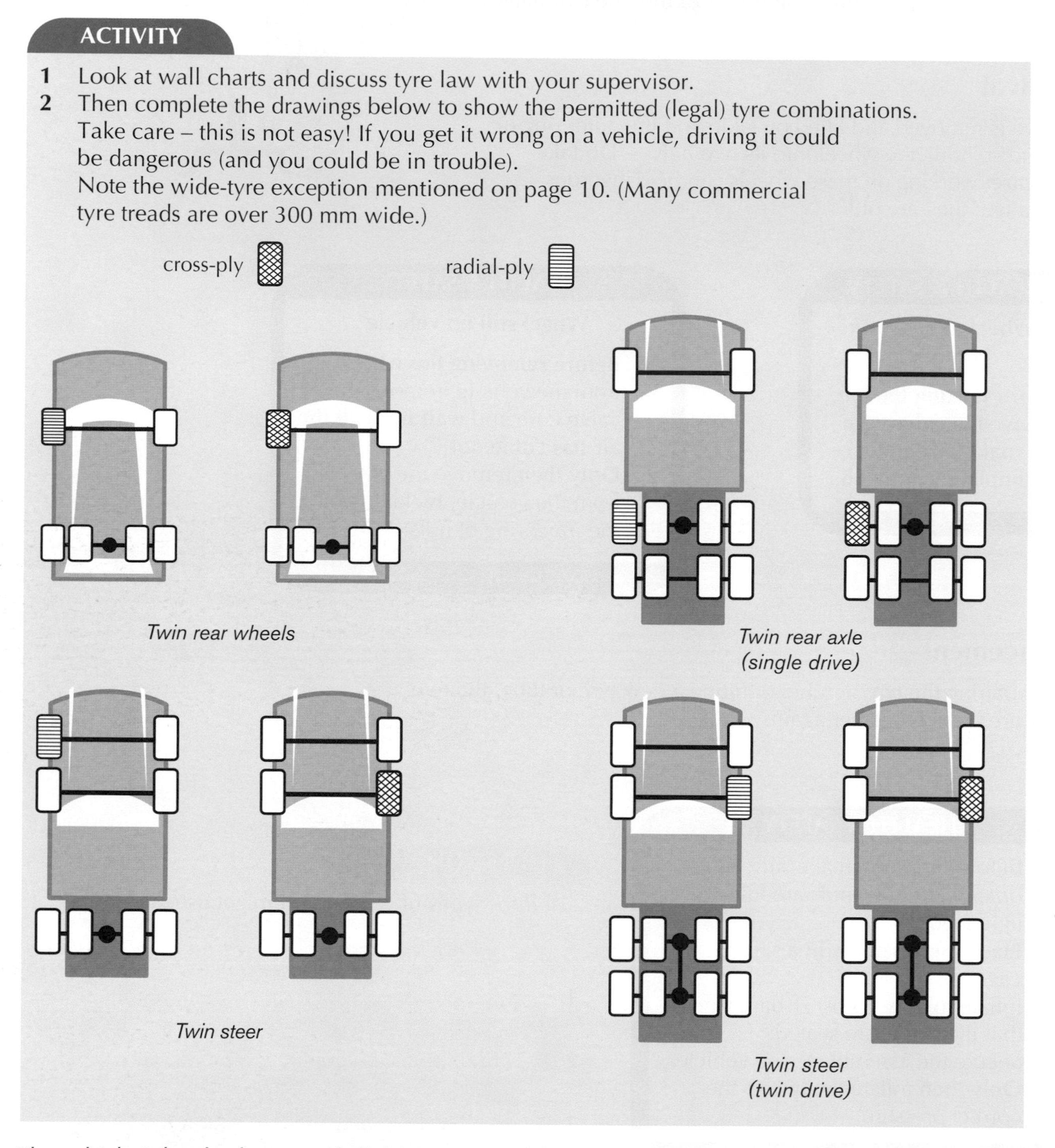

Twin rear wheels

Twin rear axle (single drive)

Twin steer

Twin steer (twin drive)

The vehicle *is* legal, of course, if all the tyres are of the same construction.

Twin wheels

Many commercial vehicles have **twin wheels** at the rear, to help carry heavy loads. In such cases, each *pair* of wheels must be evenly matched with each other. Also, the pair must match those at the other end of the same axle.

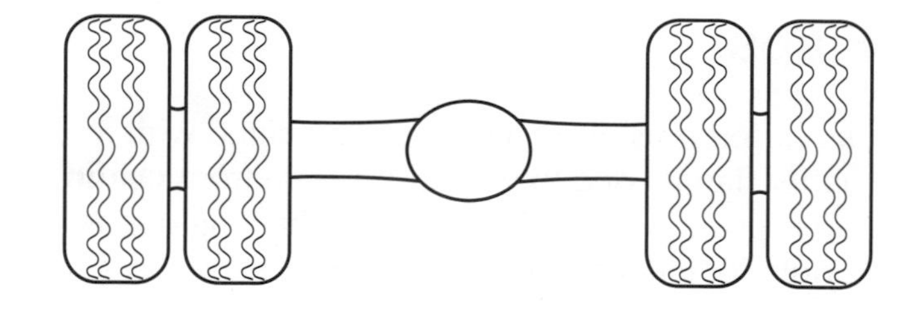

All the wheels and tyres on any one axle should be:

- the same size
- inflated to the same pressure
- worn equally
- the same type of tyre
- (preferably) the same make.

Rather than just *looking* at two tyres or trying to measure their *diameters*, the easiest way to spot differences in size is by measuring the **circumferences**. The circumference is the distance *around* the tread.

It can be measured very simply, using a **circumference tape**. The circumferences of the two tyres should be equal within ½ inch (12 mm).

QUESTION

The illustration shows twin tyres of slightly different sizes. What might be the results in use?

Load ....

Wear rate ....

Chance of blow-out...........................

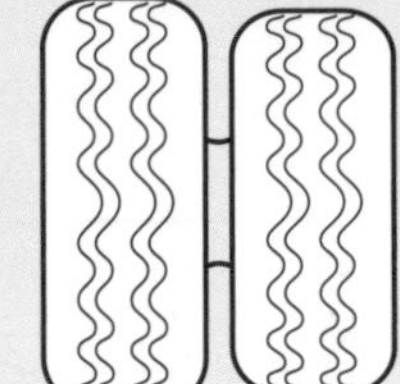

Wheel security

Securing a wheel to a vehicle is one of the most important jobs you will do as a tyre technician. **If you do not fix the wheel properly you could cause an accident or even death!**

Securing a wheel is not simply a matter of tightening up the wheel nuts. It is not a test of your muscle power. For a wheel to be properly secure, two things must happen:

1. The wheel must be **correctly located** (positioned) on the hub.
2. The wheel nuts must be tightened to the **correct torque**. In other words you must apply just the right amount of turning force with the wheel spanner.

Don't be *too* concerned: neither of these is difficult, given a little care and practice.

Wheel location

On most cars and light commercial vehicles, locating the wheel is quite simple. There is a **taper** on the wheel nuts, and this matches a taper on the wheel. As the nuts are tightened, the tapers position the wheel correctly.

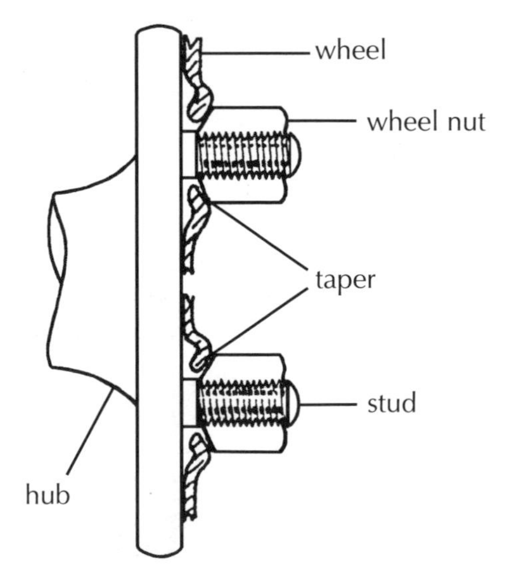

QUESTION

If the wheel nuts shown on the drawing were turned round, would the wheel be properly located?

☐ yes ☐ no

ACTIVITY

Look carefully at wheel nuts from several vehicles. Notice where they seat on the wheels. Apart from size, are they all the same?

☐ yes ☐ no

No doubt you soon found that there are several types of wheel fixings.

Shown here are several typical wheel fixings, as fitted to commercial vehicles using twin wheels.

HEALTH & SAFETY

Don't mix different types of wheel nuts, even if they fit onto the studs – the wheels might come loose.

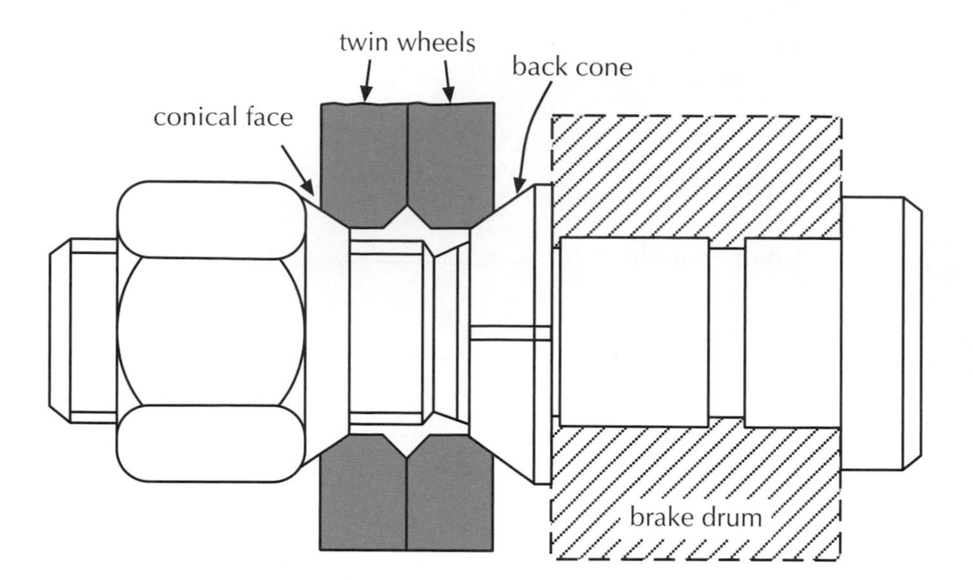

Typical British Standard. Wheel location depends on the conical faces of the nut and the back cone

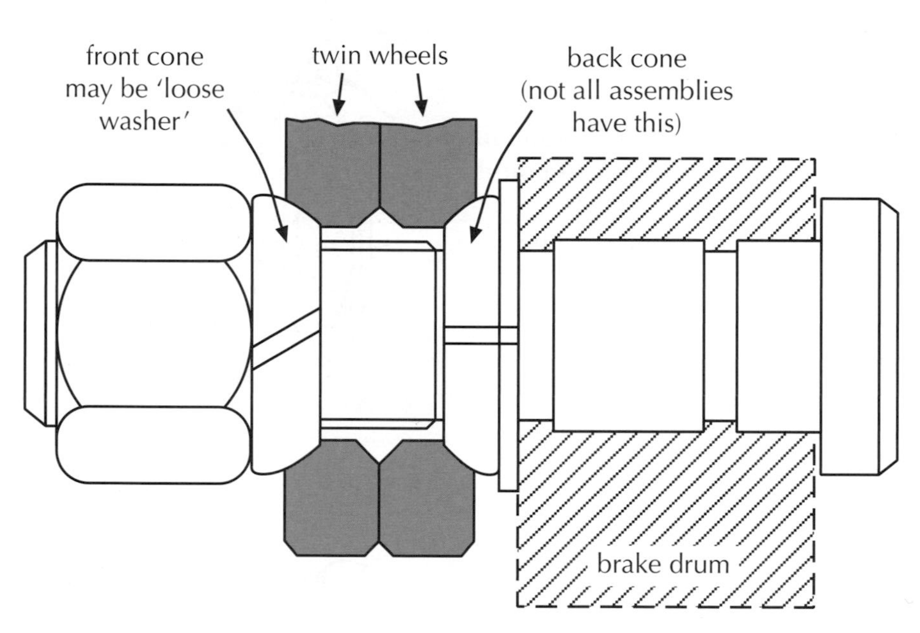

DIN-type spherical fixing. Similar to the British Standard, but location depends on the spherical (rounded) faces of the nut and the back cone

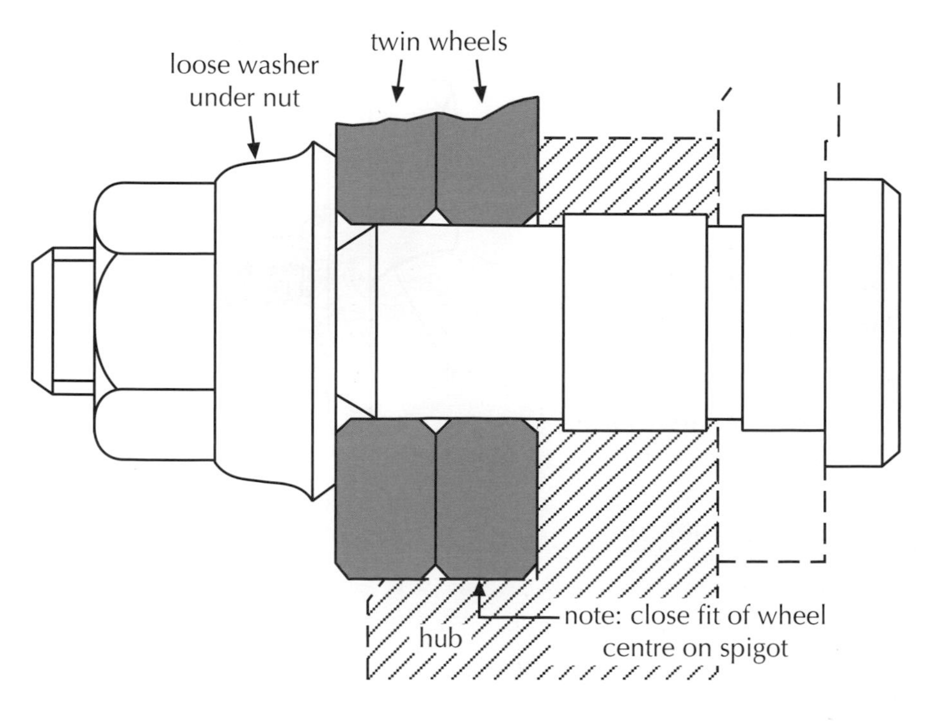

Volvo-type spigot mount. Wheels must have a very accurately machined centre bore. This slides onto an accurately machined spigot which sticks out of the hub. Note: the ISO-type spigot mounting is similar to the Volvo type

With spigot-type fixings it is the hub that *positions* the wheels: what the wheel nuts do is clamp everything in place. The loose washers reduce wheel nut friction. This fixing gives very good results in use.

How tight?

First, the wheel *must* be correctly located. Depending on the fixing, the position of the wheel will be set by:

- gentle pressure on the wheel nut cones (with the nuts lightly tightened), or
- the centre spigot and the wheel studs.

The *turning force* applied to a nut is properly called the **torque**. You can find out the correct torque for each nut from a wall chart in the workshop. In the UK torque is usually shown in units of **pounds feet** (**lb/ft**). The international unit of torque, though, is the **Newton metre** (**Nm**), and this is now widely used.

Suppose that when tightening a nut you used a spanner 2 ft long and an effort on the end of it of 100 lb. What would be the torque?

$$\text{Torque} = 100 \text{ lb} \times 2 \text{ ft}$$
$$= 200 \text{ lb/ft}$$

Instead of estimating or guessing the torque applied, you should *measure* it exactly.

QUESTIONS

1 What is the name of the tool used to measure torque on nuts?

....

2 If 1 Nm = 0.72 lb/ft, how many lb/ft = 50 Nm?

....

QUESTION

Suggest *three* things on a hub spigot (or on a wheel-centre bore) that could stop the parts fitting together properly.

1 ....

2 ....

3 ....

TIP
Before refitting a wheel, make sure that all nuts, all wheel-stud hole tapers, the brake drum face and the wheel centre bore are clean. Threads should be free and lubricated with light oil.

ACTIVITY

Choose *three* very different vehicles. Look at a wall chart or vehicle handbook. Then fill in the table below.

Vehicle		Wheel location:	Wheel nut torque
Make	**Model**	**wheel nut or spigot**	(state lb/ft or Nm)

After you have tightened the wheel nuts – especially on a commercial vehicle – advise the driver to check the tightness (but not *over*-tighten) from time to time. A common recommendation is that the first check should be between 50 km and 250 km.

Regrooving and recutting

Large commercial vehicle tyres are very expensive. They are therefore designed to last as long as possible.

The tread pattern is deeper than on a car tyre. Further, there is a thick layer of tread-type rubber *under* the tread. When the pattern becomes worn, the grooves can be cut a little deeper. This is called **regrooving** or **recutting**.

HEALTH & SAFETY

Regrooved (recut) tyres must NOT be used on cars, motor cycles or light commercial vehicles.
If you have any doubts, check with your supervisor.

Requirements for regrooving

- Only tyres that are in good condition and that are marked 'REGROOVABLE'.
- The tread should be worn to about 3 mm.
- Grooves may sometimes be made no more than 3–4 mm deeper than the original. With some makes of tyre, they could be much deeper.
- Grooves must follow the original tread pattern.
- Tyres may only be regrooved once.

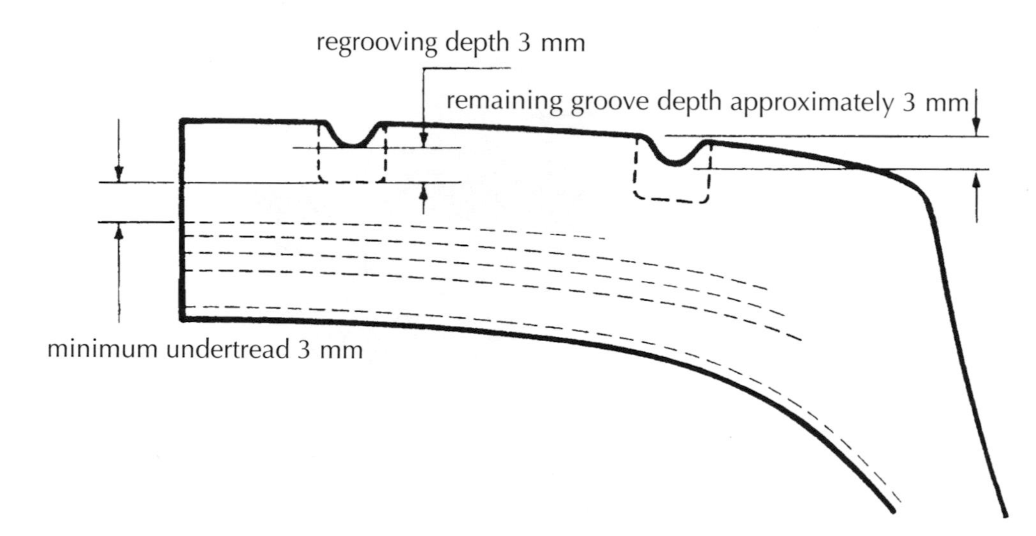

Remould and retread tyres

Remoulding means renewing almost all of the rubber on the tread and the sidewalls of a tyre. **Retreading** means renewing the tread rubber, down to the shoulders of the tyre (also known as **top-capping**).

Only tyre casings in first-class condition, apart from the worn tread, are suitable. The work is done by the tyre manufacturers or specialist firms. The products are marked 'REMOULD' or 'RETREAD'.

QUESTION

What is the main advantage of remould or retread tyres over new tyres?

....

Agricultural tyres

Agricultural tyres have a lot in common with commercial vehicle tyres. They may be of radial, cross-ply or bias-belted construction.

However, there are some big differences in the conditions in which they are used. For this reason, many special tyres are available.

Consider a tractor and trailer.

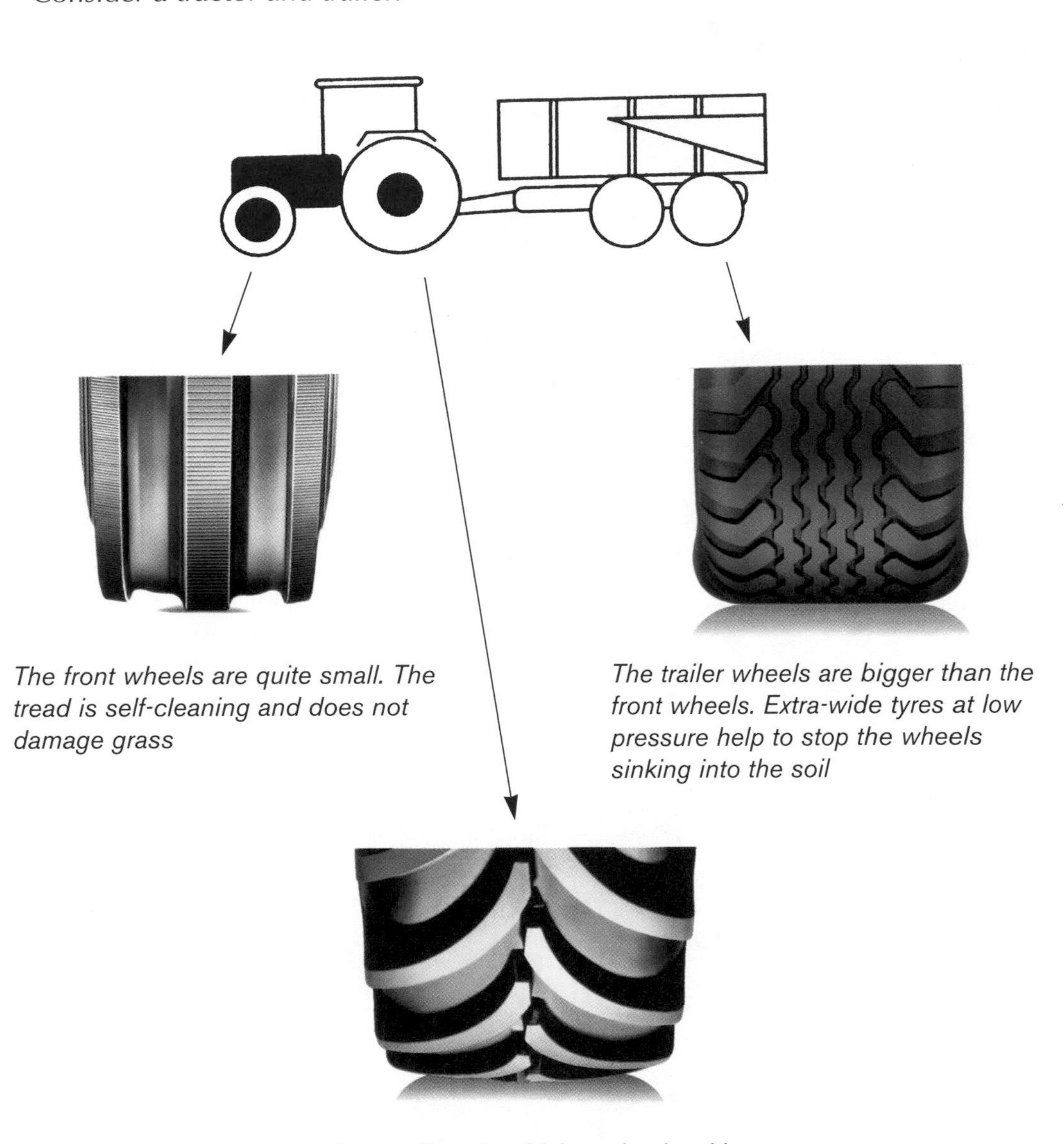

The front wheels are quite small. The tread is self-cleaning and does not damage grass

The trailer wheels are bigger than the front wheels. Extra-wide tyres at low pressure help to stop the wheels sinking into the soil

Large-diameter driving wheels with deep-tread grooves give a good grip. A wide section spreads the load, reducing the chance of sinking

QUESTION

Look at the tractor and trailer tyres shown above. Which are directional and which non-directional?

1	Front wheels	☐ directional	☐ non-directional
2	Driving wheels	☐ directional	☐ non-directional
3	Trailer wheels	☐ directional	☐ non-directional

Speed symbol and load index

The speed symbol on agricultural tyres is a letter 'A' followed by a number. This indicates the tyre's maximum permitted speed.

The **load index** number works in the same way as with car and commercial tyres.

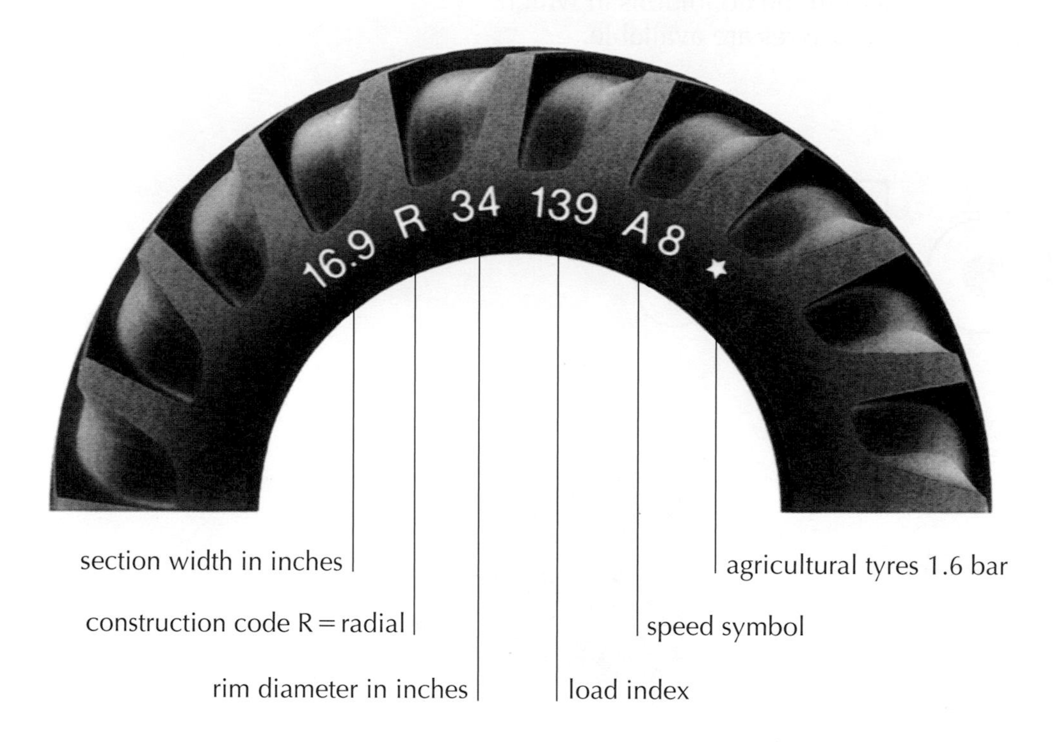

An agricultural tractor tyre, with speed symbol and load index

QUESTION

Reading from the speed index table, what is the maximum permitted speed for the tyre shown?

.... kph

Speed symbol	Max. speed (kph)
A1	5
A2	10
A3	15
A4	20
A5	25
A6	30
A7	35
A8	40

Speed index table

Liquid ballasting

One easy way to increase grip is to partly fill the driving tyres of a tractor with water. The water is **liquid ballast** (commonly a water and special antifreeze solution). One quarter (25%) of the space is left for air, so that the tyre can be inflated to the correct pressure. Note that a special **air/water tyre valve** is needed.

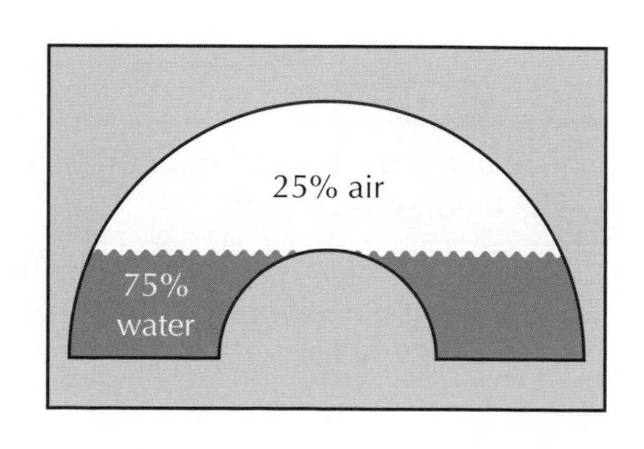

Earth-mover tyres and industrial tyres

Tyres for earth-movers are similar to tyres used on agricultural and commercial vehicles. However, they are often used in harsh conditions. For this reason they are made extra tough, to resist damage.

Earth-mover tyres may be very big indeed, and they are very expensive. They may be used in difficult conditions such as:

- underground, in mines
- in hot deserts
- in deep mud
- among sharp rocks
- fast- or slow-running.

Tyres are made to suit the work being done. Special treads, special rubber compounds and special types of construction are used.

The range of industrial tyres is extremely varied, from the small inflatable tyre for a wheelbarrow to the solid tyres on very heavy-duty works vehicles such as fork-lift trucks.

ACTIVITY

Look at manufacturers' catalogues. List *three* very different industrial tyres.

Vehicle Make/Model	**Tyre size**	**Tyre construction** (cross/radial/bias)	**Directional or non-directional?**

TIP
Scrapped tyres are 'controlled waste'. Failure to dispose of them correctly is a criminal offence, with the possibility of an unlimited fine.

Chapter 6

Electrical

In this chapter you will learn about:

- **batteries** – checking and maintenance
- **jump-starting** vehicles
- **alternators** – checking belt tension and output
- **lights** – operation and alignment

In modern vehicles the electrical systems are crucial. For example, if the battery is weak or if the alternator does not charge, then the vehicle probably will not even start.

In this chapter you will learn about batteries and alternators and how to check them. You will also learn a method of checking the vehicle's lights, and find out how to align the headlamp beams. There is also a section on starting an engine with a flat battery.

The battery

Electrical power is vital for a car to work. Just as a torch needs electricity stored in a **battery**, so a car needs electricity stored in a much larger battery. A torch battery gradually loses its power as it is used, and must then be renewed. Car batteries are much too expensive to renew, except after a long period in service. So, they are made capable of being recharged. It is the job of the **alternator** to keep recharging the battery when the engine is running.

Batteries used on cars and commercial vehicles

The battery used in a car is called a **lead acid battery**. The active plates within the battery are made from two types of lead. The liquid inside the battery is sulphuric acid, mixed with distilled (pure) water. This mixture is called the **electrolyte**.

Some batteries are *maintenance-free* (pages 92–93).

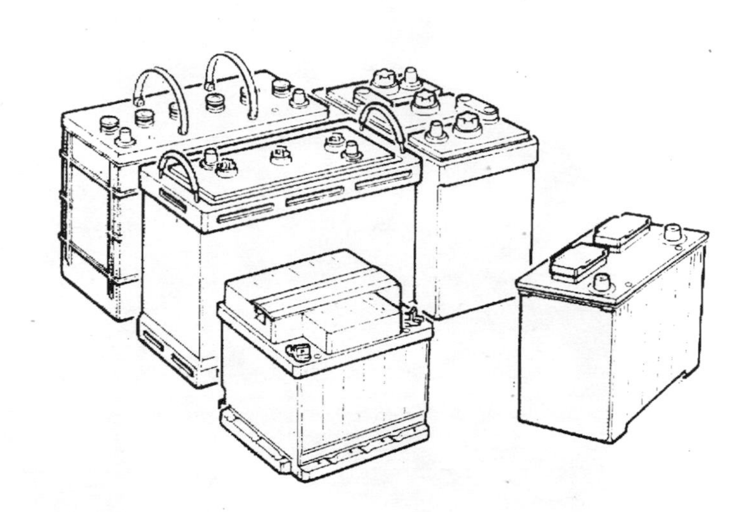

ACTIVITY

Some of the batteries shown opposite are 12 V and some are 6 V. On each battery, write the correct voltage.

Modern **electric cars** may use **alkaline batteries**.

Compared to the lead acid battery, the alkaline battery will give better performance and a longer life.

QUESTIONS

1 How many cells are there in a 12 V battery?

....

2 How many cells are there in a 6 V battery?

....

Typical 12 V battery

Note the parts of the battery.

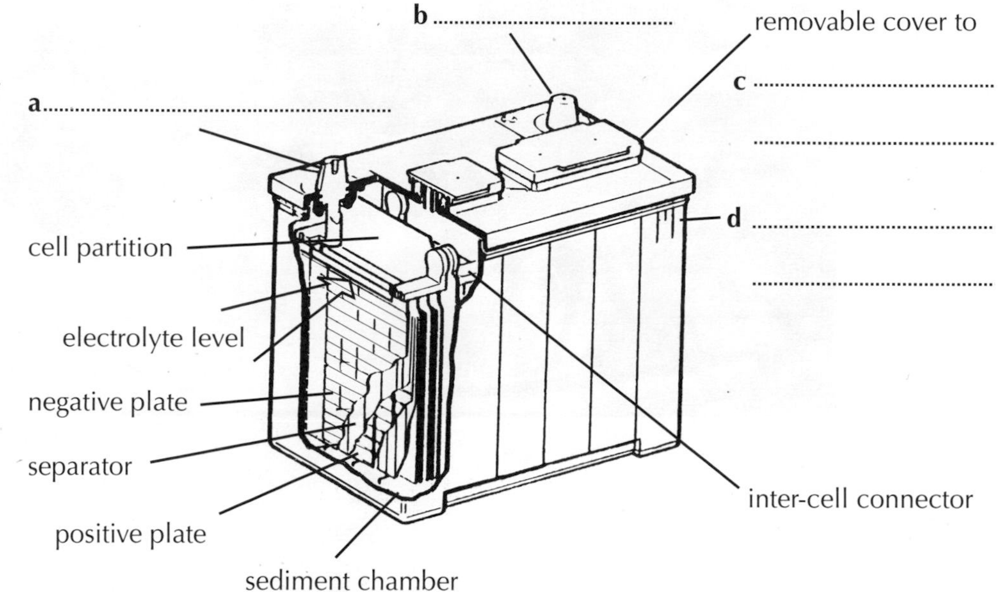

ACTIVITY

1 Fill in the missing names on the drawing opposite.
2 Examine a demonstration battery that has been cut open like this, and identify the main parts.

Battery maintenance

It is important to keep the battery in good working order. At each service, make these checks.

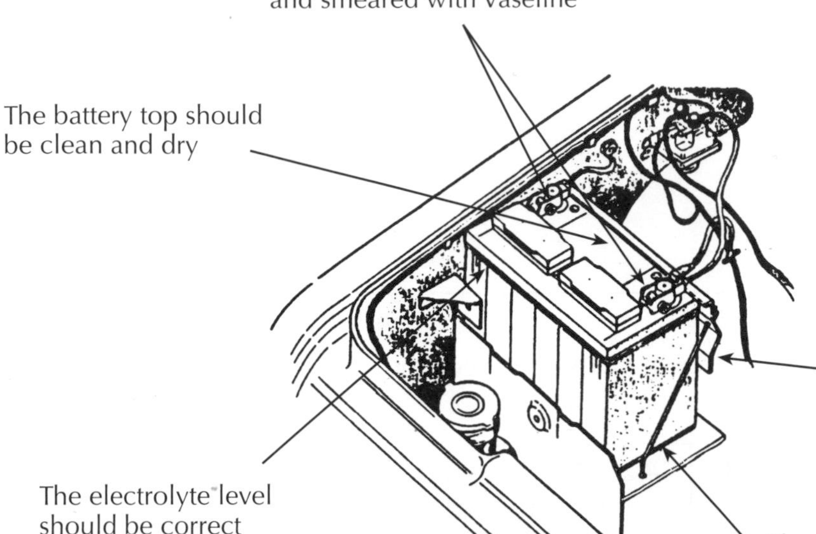

HEALTH & SAFETY

The battery must be topped up only with distilled water. **Do not overfill.**

Recharging batteries

One of the jobs you will be expected to do is to recharge batteries.

QUESTION

What is a typical low charge rate?

.... amps

WORKSHOP

Recharge a battery using a bench charger.

1. Remove the battery from the vehicle.
2. Carry the battery to the bench charger.
3. Check the level of electrolyte.
4. Charge at a low charging rate.
5. When the battery is fully charged, **switch off the battery charger before removing the leads. A spark could cause the battery to explode.**
6. Remove the charger leads from the battery.
7. Replace the charged battery in the vehicle.
8. Start the engine.

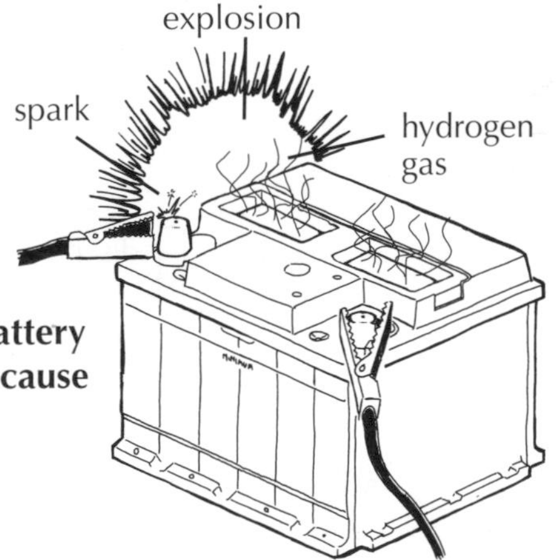

TIP

When connecting charging leads, always connect:

- charger positive to battery positive
- charger negative to battery negative.

HEALTH & SAFETY

- Wear gloves and overalls when carrying a battery.
- The acid can burn if spilt.
- Batteries are heavy, and hard to carry.
- The battery charging area must be well ventilated.

Testing the battery

To test the battery state of charge, you use a **hydrometer**. This measures the specific gravity of the electrolyte. The higher the number, the higher the state of charge.

Specific gravity	Battery state of charge
1.120	discharged
1.240	70% charged
1.280	fully charged

Specific gravity or (**relative density**) is a measure of the *weight* of the **electrolyte**. When fully charged, the battery acid (electrolyte) is heavy; when discharged, it is light.

WORKSHOP

1. Use a hydrometer (similar to the one shown below) to check the specific gravity in each cell. Write down the readings.

Cell number	1	2	3	4	5	6
Specific gravity						

2. What was the battery's state of charge? ....

3. What was the battery's condition? ....

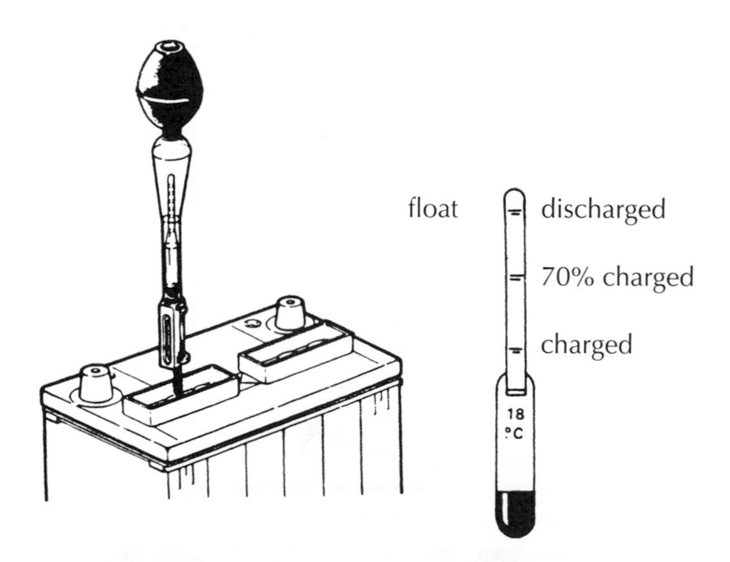

A hydrometer

TIPS

- If all readings are about the same, the battery is in good condition.
- If one cell gives a much lower reading than the others, the battery is faulty.

HEALTH & SAFETY

- Never allow the electrolyte to drip. It can burn holes in your clothes, and burn your skin.
- Wear rubber gloves.

Maintenance-free battery

A **maintenance-free battery** is completely sealed. To check the state of charge, look down the indicator hole on top of the battery. Inside is a built-in hydrometer.

Colour	Battery state of charge
Green dot	above 65% charged
Dark or black dot	below 65% charged
Clear or yellow	low electrolyte or battery fault

WORKSHOP

Look at a maintenance-free battery. Check its state of charge.

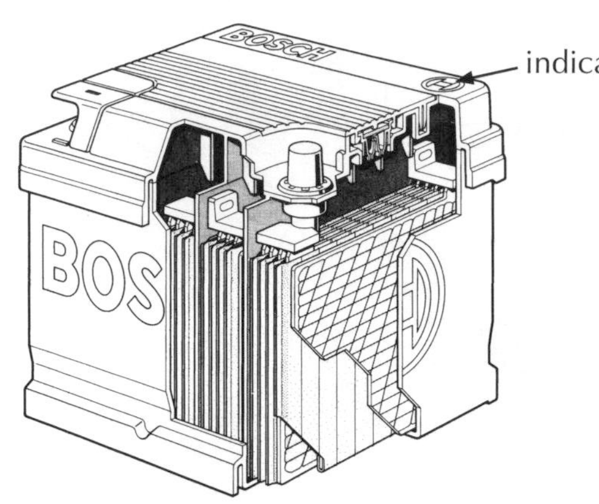

HEALTH & SAFETY

- Make sure no metal object falls across the battery terminals. It could cause an electric arc. The metal in contact with it might heat up rapidly, and that could cause a fire.
- Never wear rings or wrist jewellery while working. They could short across the battery terminals, causing an electrical burn.

Jump-starting a car engine

When a car won't start because the battery is 'flat', another battery and **jump leads** may be used to start the engine. The leads must be connected in the right order, as shown opposite.

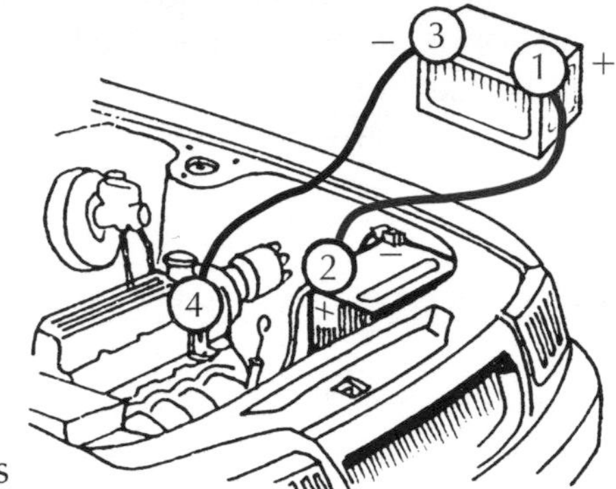

Connecting

First lead: connect the working battery's *positive*, **1**, to the flat battery's *positive*, **2**.
Second lead: connect the working battery's *negative*, **3**, to the car's *engine*, **4**. Clamp the lead to a metal part. The clamp should be as far away from the flat battery as possible.

Disconnecting

After starting the car, remove the leads in the reverse order.

QUESTIONS

1 The instrument used to measure the specific gravity of the electrolyte in a battery is:
a a thermometer?
b a hydrometer?
c a voltmeter?
d a high rate discharge tester?

....

2 If the electrolyte level is low, the battery should be topped up with:
a diluted acid?
b clean distilled water?
c electrolyte with a relative density of 1.280?
d pure bottled mineral water?

....

3 After removing corrosion and cleaning the battery terminals, they should be:
a polished and kept dry?
b smeared with silicone grease?
c smeared with petroleum jelly?
d painted?

....

4 When a battery is being charged on the bench it gives off hydrogen freely. When is it most dangerous?
a when in a very discharged state, being put on charge?
b when being brought into service and given its first filling of electrolyte?
c when it needs topping up with acid as well as distilled water?
d when being disconnected from the charger?

....

The alternator

What does the alternator do?

When the engine is running, the **alternator** supplies current to the car's electrical system. It also charges the battery.

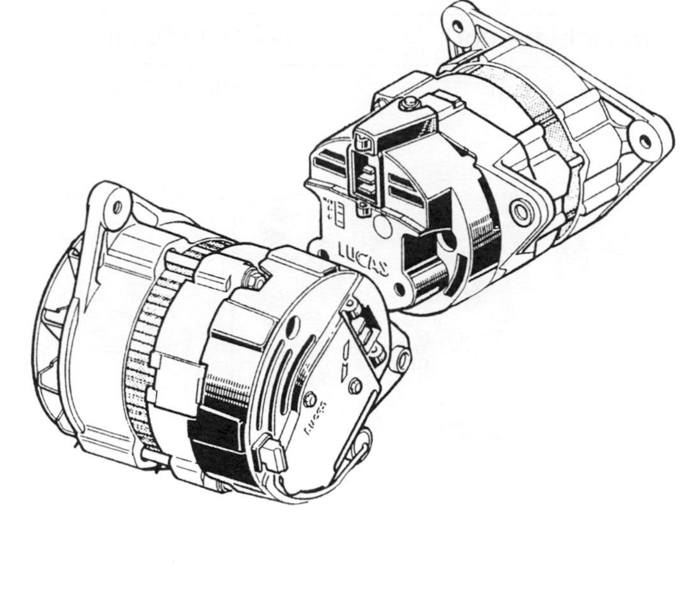

When does it need maintenance?

Alternators need very little maintenance.

The most common fault is a slack (or slipping) **drive belt**. This shows up as:

- an apparently weak battery, and/or
- a squealing noise – usually just after the engine is started from cold.

Before testing an alternator, check:

- the battery condition and its state of charge
- the connections between the battery and the alternator
- for breaks in the leads
- for a loose or worn drive belt.

Adjusting the alternator belt

1 Slacken bolt 1, and then bolt 2.
2 Pull the alternator firmly outwards at the bottom, to make the belt tight. Hold it in place and tighten bolt 2.
3 Check the tension.
4 If it is correct, tighten bolt 1. If not, slacken bolt 2 again and repeat step **2**.

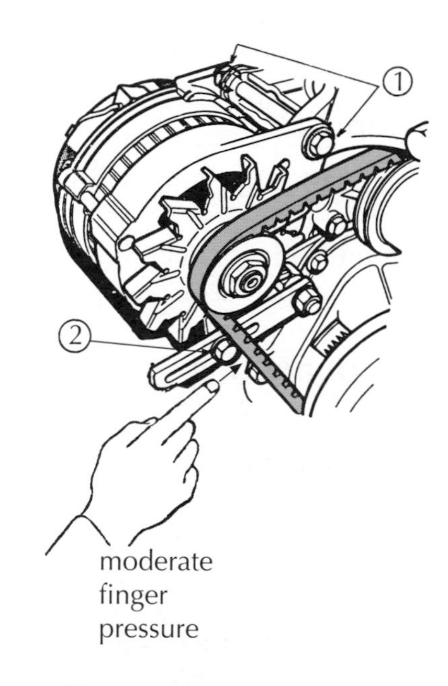

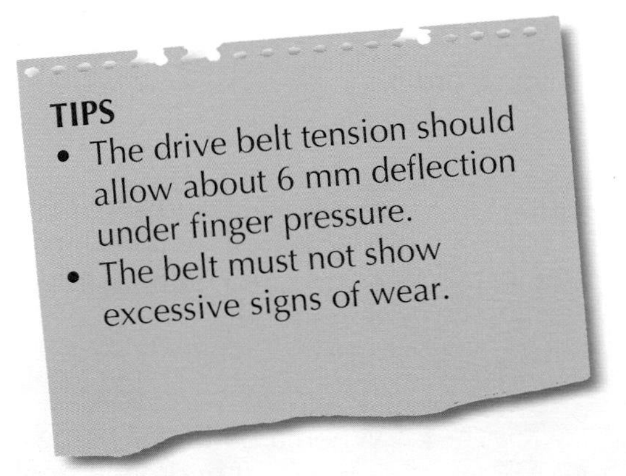

WORKSHOP
Measure (and if necessary adjust) the drive belt tension. Assess the belt movement as shown.
The drive belt tension was: ....

Checking the operation of the alternator

To check the alternator, use a **0–20 V voltmeter**. This meter provides a simple check, and can measure the condition of the battery at the same time.

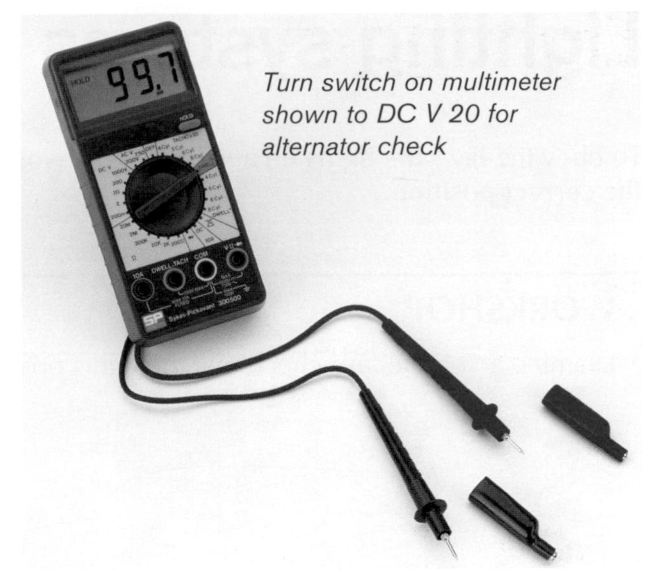

Turn switch on multimeter shown to DC V 20 for alternator check

Checking the battery

To test the battery, the engine should be *stopped*.

1 Connect the black clip of the voltmeter to the negative (–) battery terminal.
2 Connect the red clip to the positive (+) battery terminal.

If the battery is working properly, the voltage should be 12.5 V or more.

> **WORKSHOP**
>
> Check the voltage of a battery.
>
> Actual reading: ....
>
> Condition of battery: ....

> **TIP**
>
> To ensure the alternator gives a maximum battery charge output reading, switch on electrical systems such as the lights and the heated rear windscreen before starting the test. Leave them on for 3–5 minutes. (Do *not* operate the windscreen wipers – working them when the screen is dry may scratch the screen.)
>
> Using electrical systems uses up any charge that the alternator may just have given to the battery.

Checking the alternator

> **WORKSHOP**
>
> Connect the voltmeter, as with the battery check. With the engine *running* at 2000 rpm or slightly higher, the voltage should be above the battery test reading, but not above 15 V.
>
> Actual voltage reading: ....
>
> Condition of alternator: ....

Quick-check tester

Many service centres use a **'quick check' tester** to check the condition of the alternator, like the ones shown here. In these, lights show the voltage.

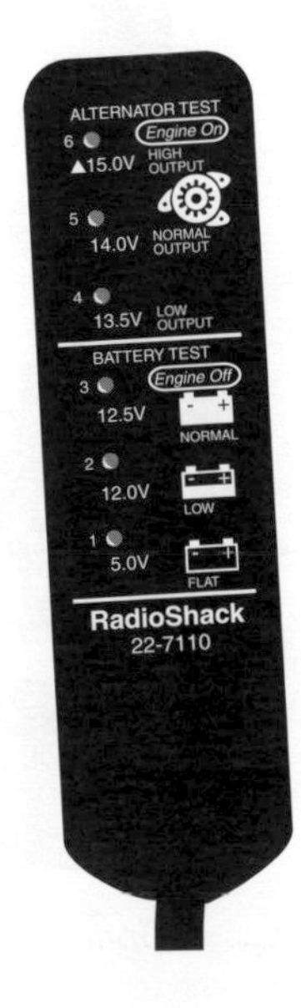

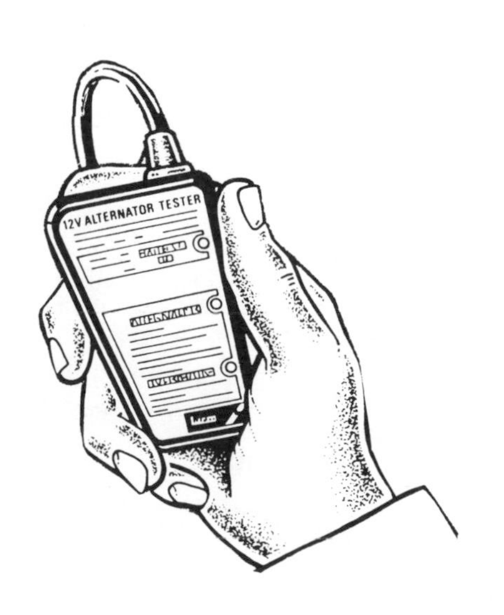

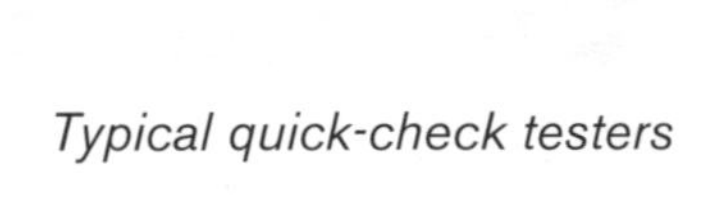

Typical quick-check testers

Lighting systems

To obey the law, **the lights on a vehicle must work properly and be in the correct position.**

WORKSHOP

Examine a vehicle and check that the lights operate properly.

Rear direction indicator check

Vehicle: Make .... Model ....

Front sidelights	☐	Brake lights	☐
Rear sidelights	☐	No. plate light	☐
Direction indicators	☐	Rear fog light(s)	☐
Headlamp main beam	☐	Reversing light(s)	☐
Headlamp dip beam	☐	Reflectors	☐

TIP
If a single light has failed it is likely that a bulb has blown. If both brake lights have failed the problem could be that the switch is faulty or the fuse has blown.

Measuring the headlamps' alignment and dip

Headlamps must be correctly **aligned** so that they give good light. In the UK they must dip to the left, so as not to dazzle oncoming traffic.

WORKSHOP

Use an **optical beam setter** to check a vehicle's headlamp aim and dip. Make sure you know what to do. Read the instructions or ask your supervisor.

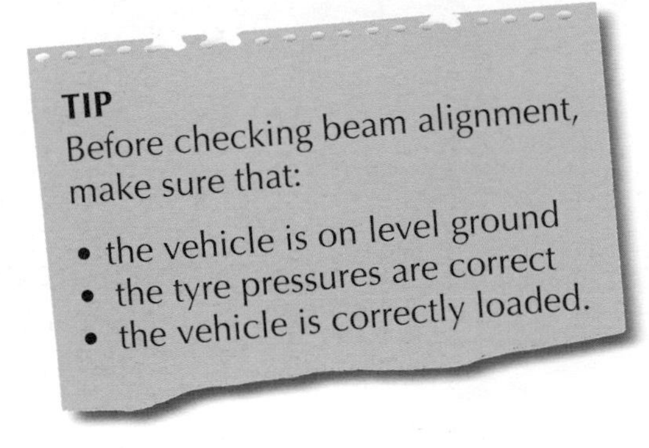
TIP
Before checking beam alignment, make sure that:

- the vehicle is on level ground
- the tyre pressures are correct
- the vehicle is correctly loaded.

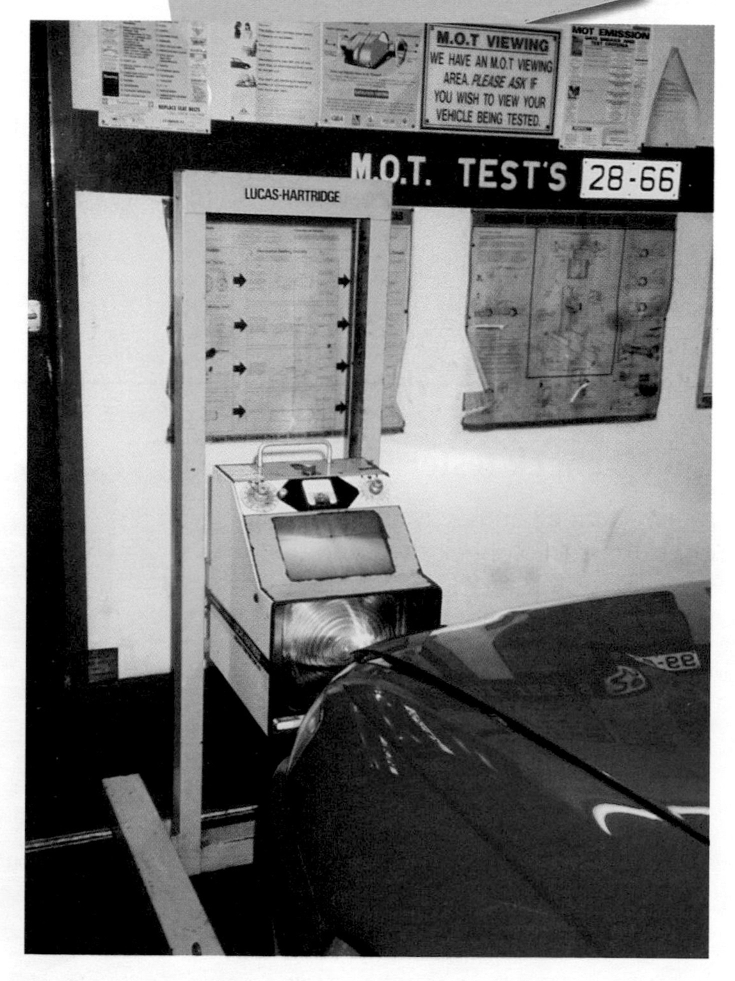

Optical beam setter

Type of beam

Most dip beams are *asymmetric*, as shown on the diagram below. In each case:

- the boundary line between high and low light intensity areas must be at least 0.5° below the horizontal line
- the break point, where the boundary line turns up at 15°, must be to the left of the vertical line.

TIP
Before starting the check, find out where the adjusting screws are on the headlamps and what each screw alters.

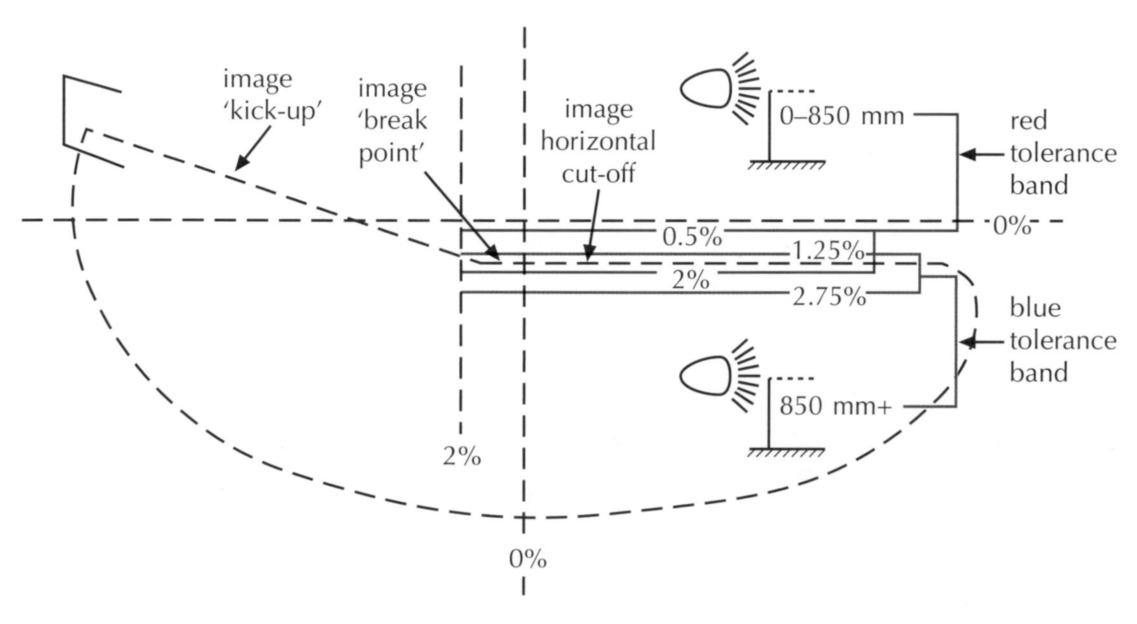

Correct light pattern for European headlamp on dipped beam (UK standards)

Fuses and relays

Should an excessive electric current flow, due to a fault in the circuit, the fuse is designed to break and so protect that circuit. Relays are electrically operated switches.

TIP
Never renew a fuse more than once without tracing the reason why the fuse failed.

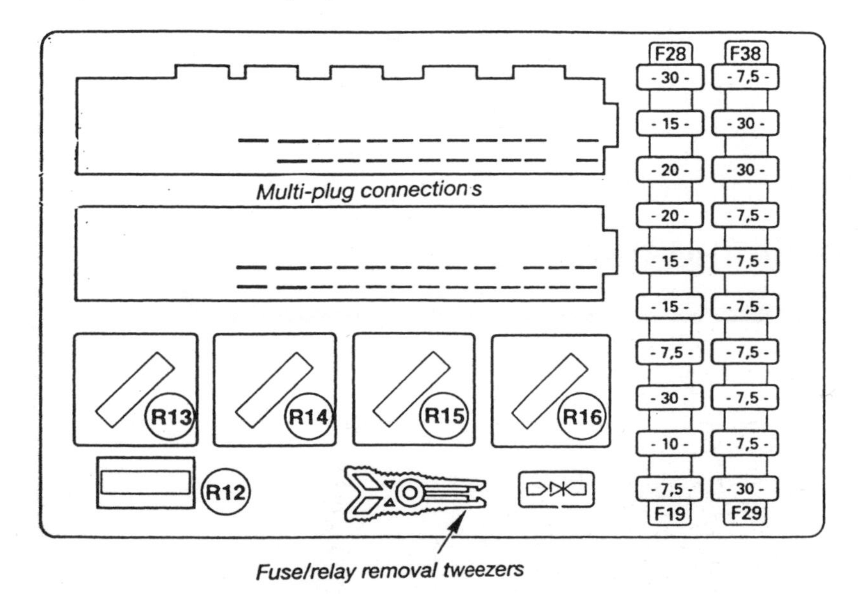

Typical Main Fuse Box Layout (Ford).
Note the different fuse ratings. (R = Relay)

Good fuse

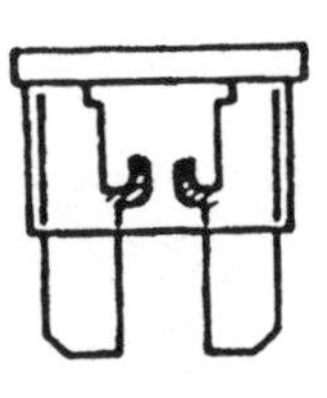

Blown fuse

Chapter 7

Brakes, suspension, steering

In this chapter you will learn about:

- **brakes** – disc brakes and drum brakes, and the hydraulic systems that operate them
- **suspension** – springs and shock absorbers
- **steering** – different steering systems, and wheel alignment and tracking

Brakes, suspension and steering are all vital for safe vehicle control and comfort. Most parts of these systems are hidden from view. Many now use advanced technology.

This chapter explains these systems and how to maintain them.

The braking system

On a bicycle you use the brakes to slow down or stop. You also use them to control the speed, as when going downhill. The brakes on a car serve the same purposes. Instead of a hand lever, the driver uses a foot pedal.

The main parts of a typical braking system are shown below. This is a **hydraulic system**, operated by the footbrake. Notice that the brake units are connected by pipes.

QUESTION

On most cars there is also a handbrake lever. When would the driver use this?

....

....

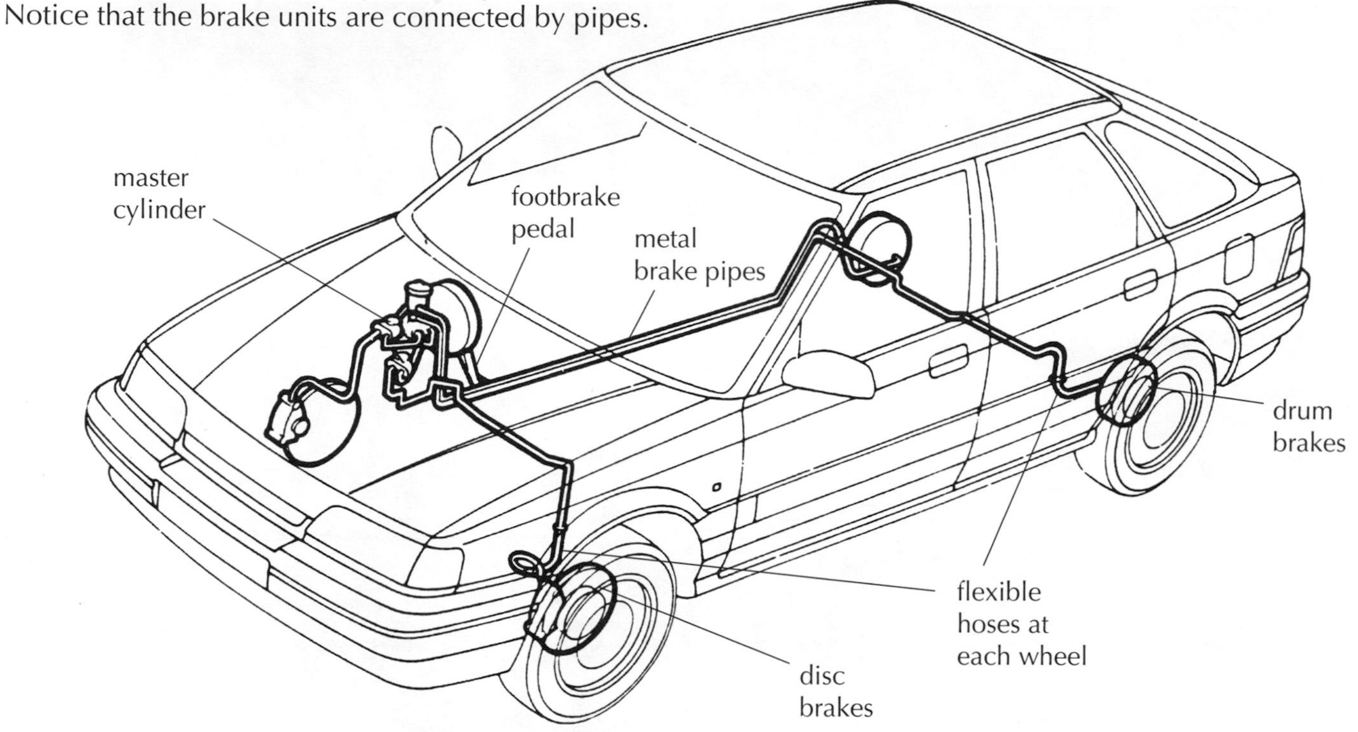

brake drums turn with the road wheels and hubs

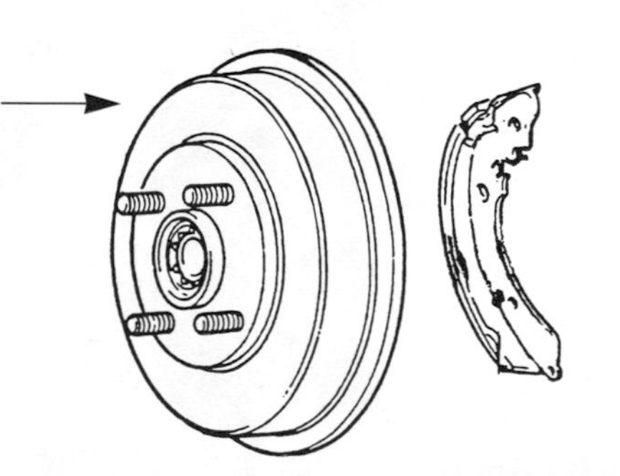

brake shoes are pressed against the drums to slow them down

brake discs turn with the road wheels and hubs

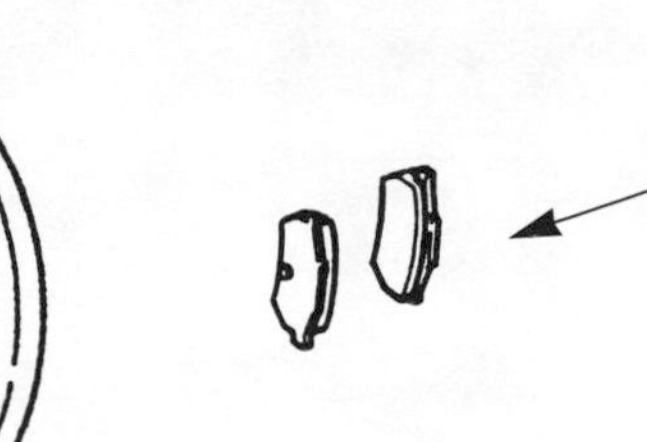

brake pads are pressed against the discs to slow them down

WORKSHOP

Examine some actual parts, as shown above. Notice where they rub against each other. This is what makes the brakes work.

Hydraulic system

Drum brakes

When you press the foot pedal, fluid is pumped along pipes from the **master cylinder** into the **wheel cylinders**. The wheel cylinder pistons force the brake shoes against the drum.

QUESTION

This is a *hydraulic* action. Why are the rubber seals needed?

....

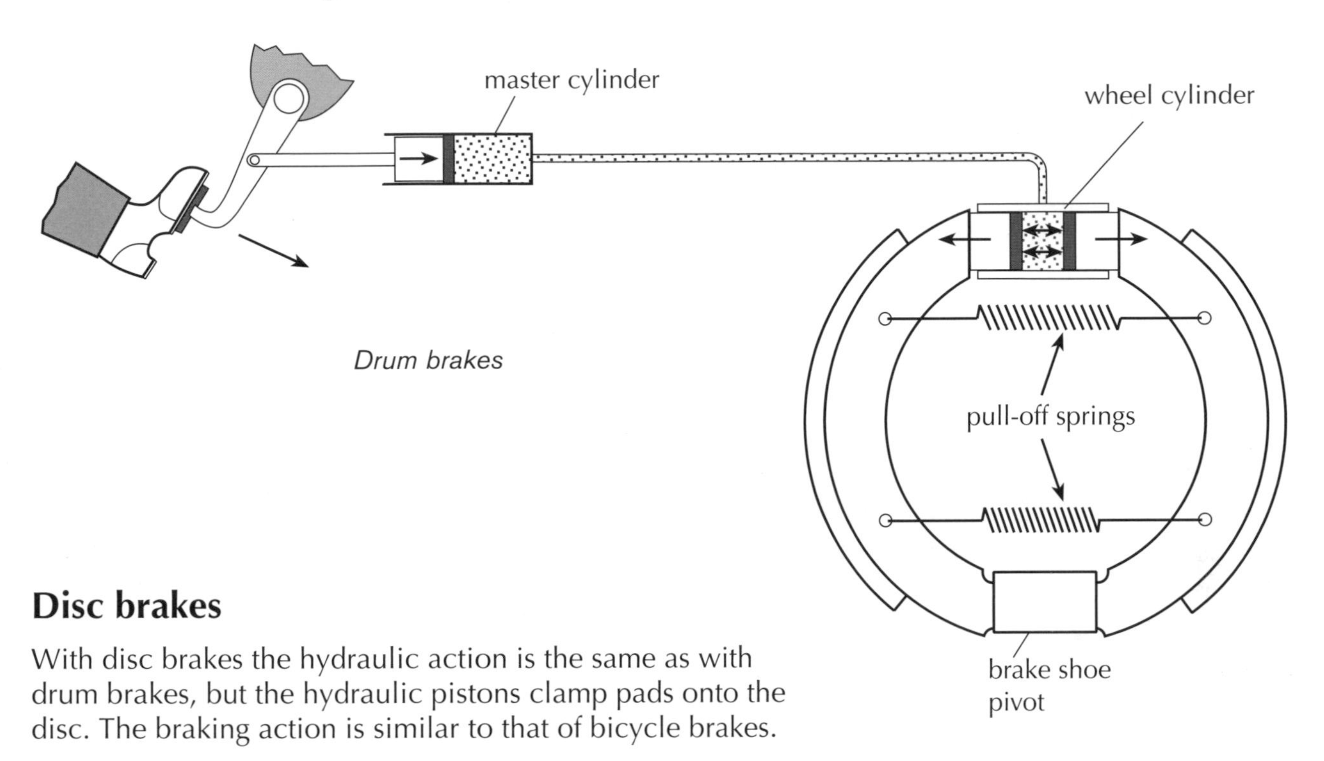

Drum brakes

Disc brakes

With disc brakes the hydraulic action is the same as with drum brakes, but the hydraulic pistons clamp pads onto the disc. The braking action is similar to that of bicycle brakes.

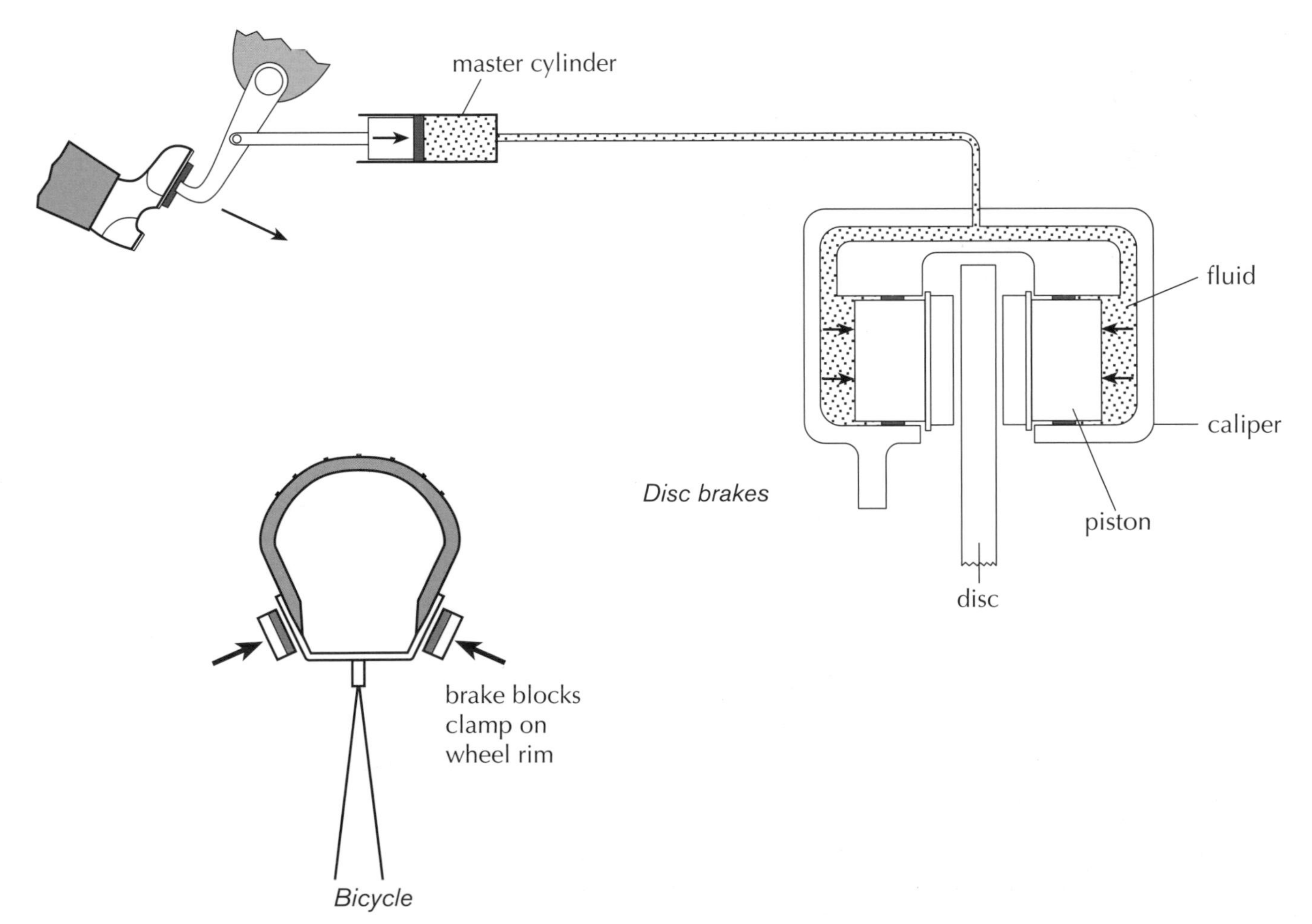

Disc brakes

Bicycle

Parking brake (handbrake)

On most cars, the parking brake works only on the rear wheels. The hand lever is joined to a cable which works the brakes. This is a **mechanical action.**

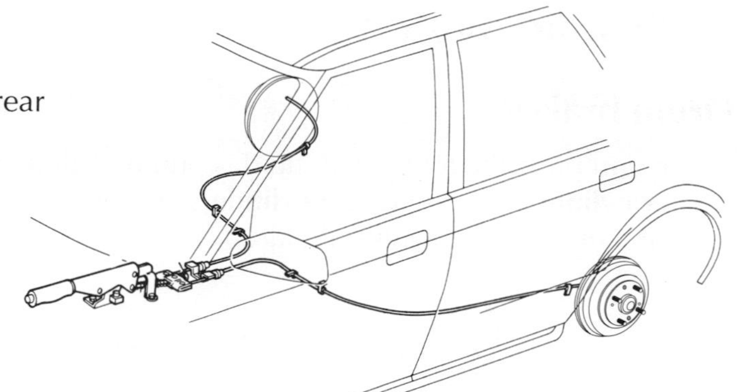

WORKSHOP

Look at a vehicle that is safely raised and has a rear brake drum removed.

Ask someone else to operate the handbrake. Watch how the brake shoes are moved by the mechanism.

Look under the car, and watch how the cables and linkage move.

QUESTIONS

1 Look at the drawing of the drum brake below. Add letters to match the names of the main parts shown.

A parking brake lever
B auto-adjuster
C return spring
D wheel cylinder
E hold-down spring

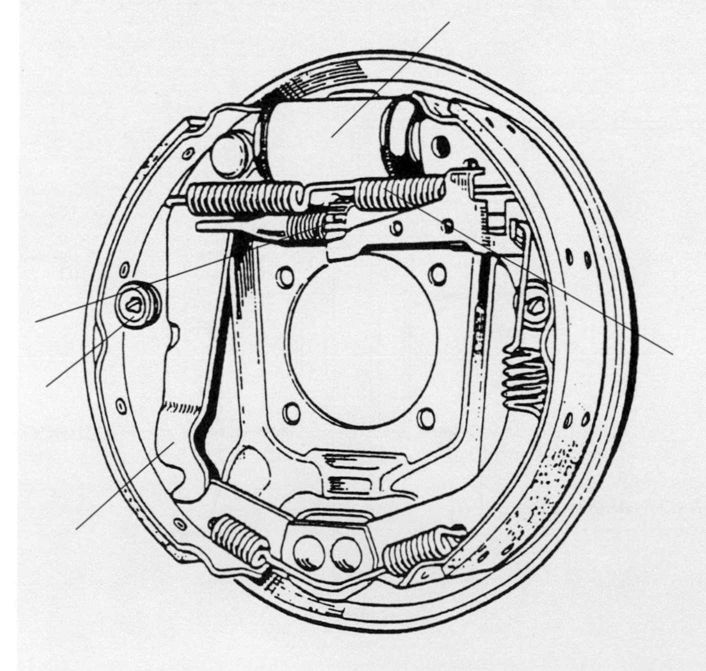

2 Look at the drawing of the disc brake below. Add letters to match the names of the main parts.

F caliper
G hydraulic pistons
I pads
J disc
K pad-retaining pins

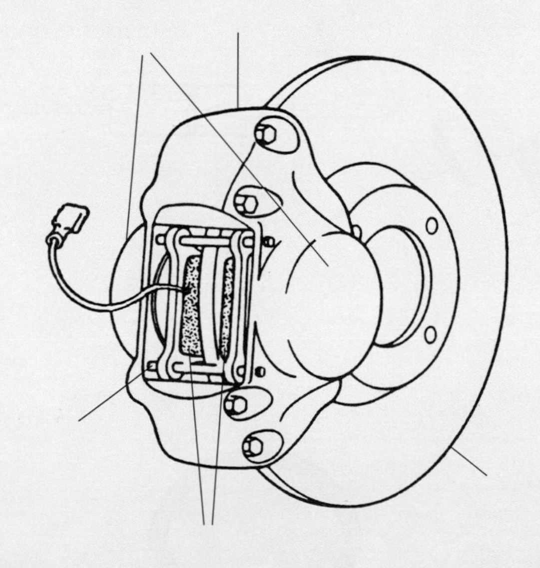

3 On the disc brake there is a cable attached to one of the pads. What is it for?

....

....

....

Many modern cars have disc brakes fitted to all four wheels. Look at car magazines or brochures to see which cars have discs all round.

Service and repair

To keep the braking system in good working order, you will need to inspect, test, adjust, repair and replace parts of the system.

QUESTIONS

Look at the drawings below. What maintenance checks do they suggest? Write your answers in the boxes.

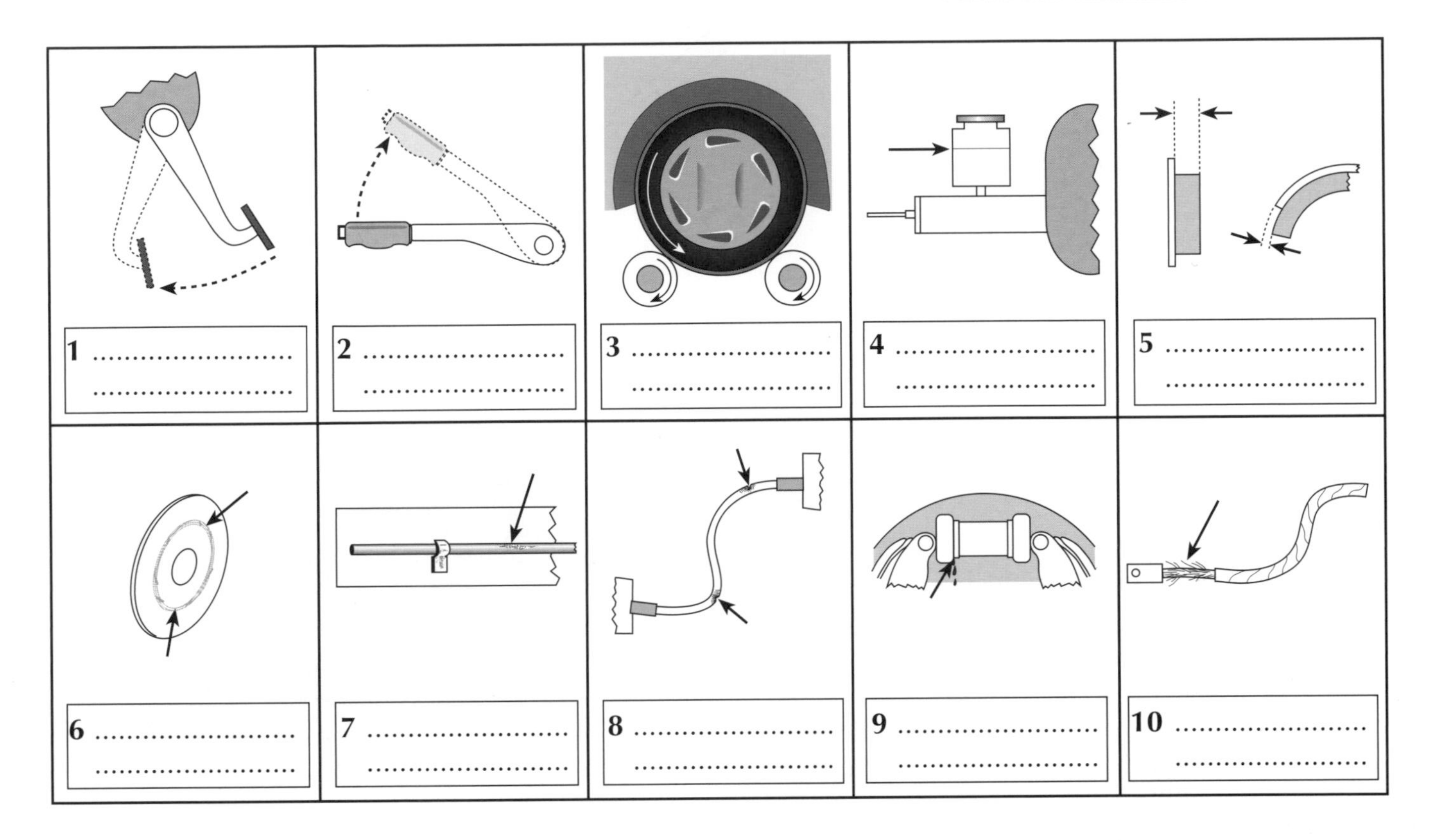

Brake adjustment

It is important to keep the **brake shoe** or **brake pad** material in the right position – just clear of the drum or disc. Without adjustment, footbrake and handbrake movement (travel) becomes too great. Most **brake adjusters** work automatically. Some vehicles, though, have *manual* adjusters at the rear. These need to be adjusted during servicing.

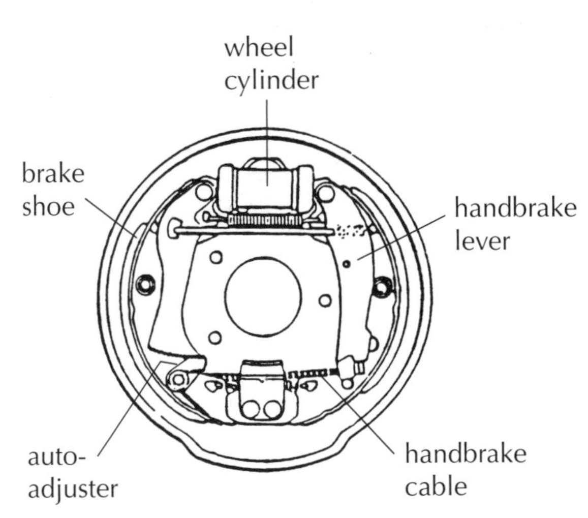

Brake bleeding

Sometimes air gets into the hydraulic system.

QUESTION

When could air get into the hydraulic system?

....

The effects of air in the hydraulic system could be:

- a 'spongy' feel when the footbrake is applied
- no braking when the footbrake is applied – the pedal just hits the floor.

The effect depends on how much air is in the system.

WORKSHOP

Feel the difference in footbrake operation with and without air in the system.

Find out how to bleed brakes safely.

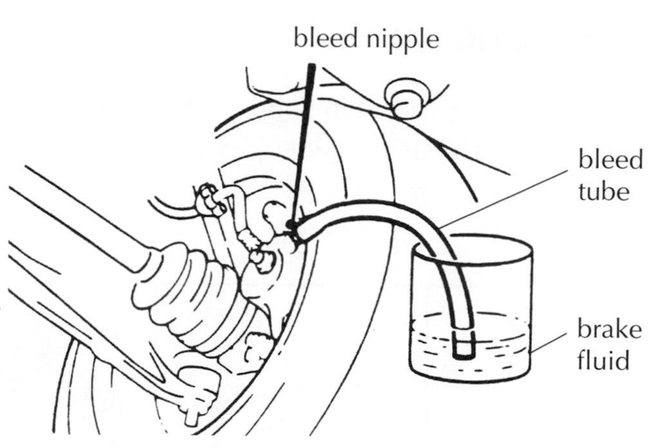

The process of forcing the air out is called **bleeding**. There are special **bleed valves** (nipples) in the calipers and the wheel cylinders. These can be released, allowing air to be pumped out of the system.

Brake seizing

Parts of the braking system can become tight or **seized** (stuck). This stops the brakes working properly. The parts that most commonly seize up are:

- the wheel cylinder and the caliper pistons
- the caliper slide mechanism
- the pads in their guides
- the adjusters
- the handbrake cables and mechanism.

TIPS

When you replace flexible brake hoses:

- Check that the pipe unions are clean.
- When the unions are finally locked in position, check that the pipe is not twisted.
- Check that the pipe does not foul other components during suspension and steering movement.

TIP

Remember: there is no point in fitting new pads or shoes if the mechanism is seized up.

HEALTH & SAFETY

- Take great care when jacking, supporting, and chocking the vehicle.
- Always use the correct tools.
- Use an extractor to remove brake dust.

QUESTIONS

1 When you release the brakes, which part moves the brake shoes away from the drums?

....

2 Which part locks the handbrake on?

....

3 Name one way of checking brake efficiency.

....

4 What is the legal minimum efficiency for foot and handbrake operation?

footbrake: .... %   handbrake: .... %

TIP

- Do not allow brake fluid to touch the vehicle's paintwork.
- Do not allow friction materials (pads/linings) to come into contact with oil or fluid, e.g. from hands or rags.

QUESTIONS

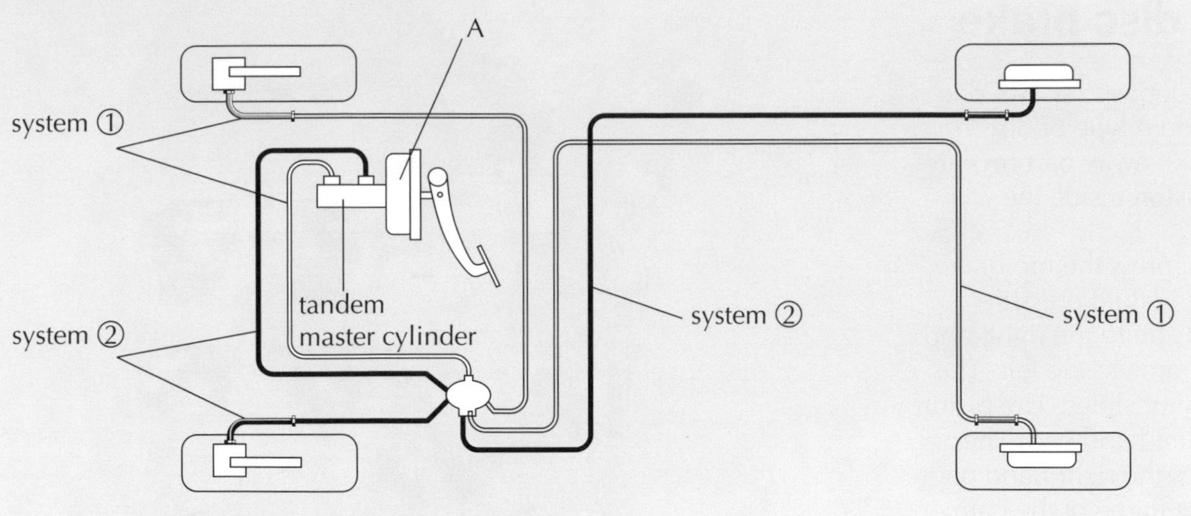

Split-line braking system

1 Look at the diagram of a split-line system shown above. Suppose one hydraulic pipe leaked: what would happen?

....

....

2 Part **A** on the drawing makes it easier for the driver to press the footbrake pedal. Name this part.

....

3 One braking system (not shown here) prevents wheel lock or skidding. This gives better, more controlled braking, especially when the road is wet or icy. What is the name of this system?

....

There are many different types of drum and disc brakes. Calipers, wheel cylinders, adjusters, handbrake systems and so on may be quite different for each of these. As you work under supervision on a variety of vehicles, you will gradually build the skill and knowledge to do this work safely.

TIP
Brakes can be quite complicated. Always strip and finish reassembling one side before you start the other. Then, if you have a problem with the reassembly, you can always look at the other side.

Sliding or floating caliper disc brake

The picture opposite shows one popular modern type of disc brake. The cut-away part reveals the **single piston** inside the cylinder.

When you press the footbrake, the increase in fluid pressure forces the piston to the right, and the caliper frame to the left. The frame 'floats' or slides. The piston applies the left-hand pad. The frame applies the right-hand pad.

The sliding frame of the caliper may become tight, or seize up. When you renew the brake pads, always check that the frame moves freely.

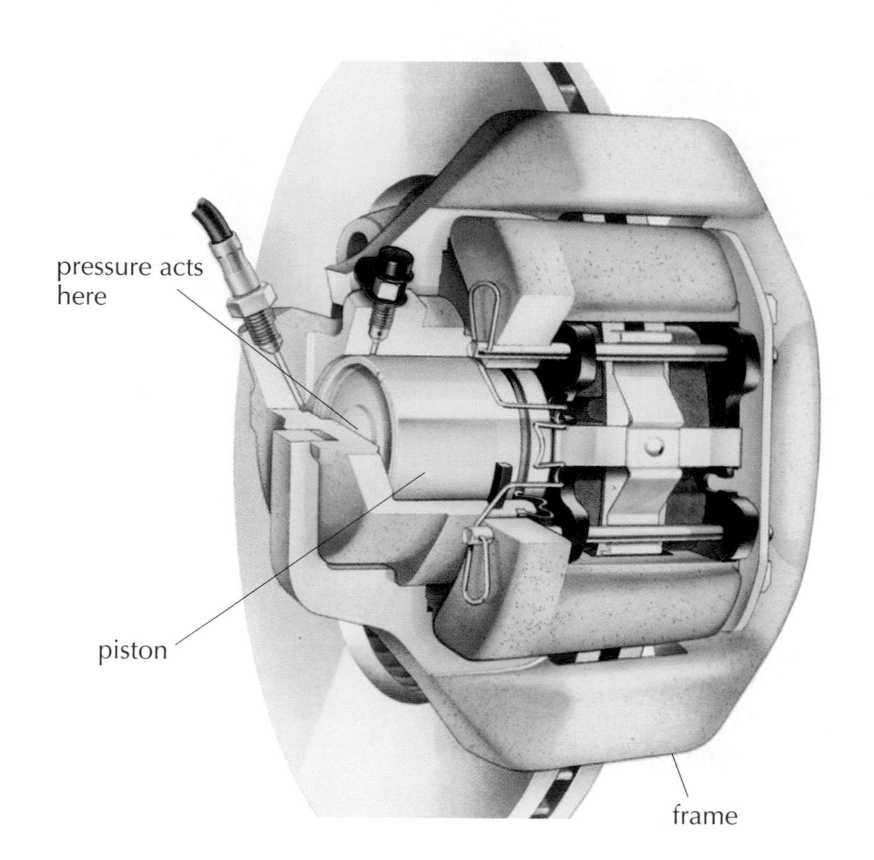

Suspension

As a vehicle travels along the road, its wheels and tyres are in direct contact with the road. They ride up over every bump, and drop into every pothole.

All these jolts would give a very uncomfortable ride, but for the **suspension system**. This cushions the chassis and body of the vehicle from what is happening to the road wheels. **Springs** between the axle and the chassis absorb the shock as the wheel rolls over a bump.

ACTIVITY

Watch some vehicles travelling along a bumpy stretch of road. Notice the up and down movement of the wheels. Notice too how much less the body moves.

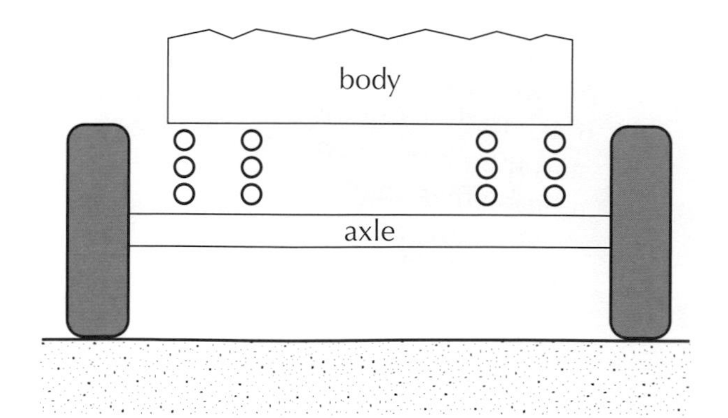

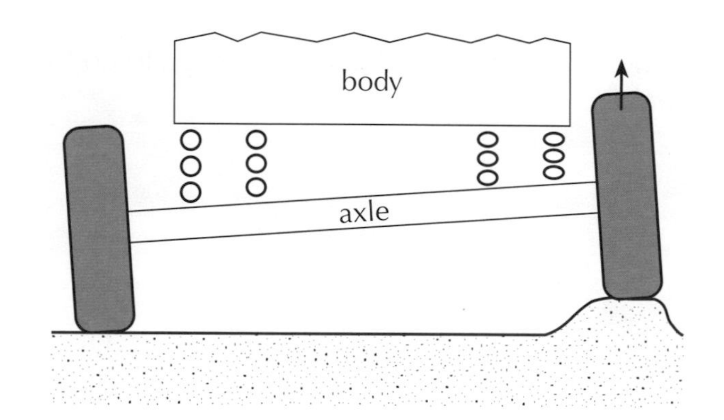

There are many different types of suspension system. Here are just some of the springs that are used.

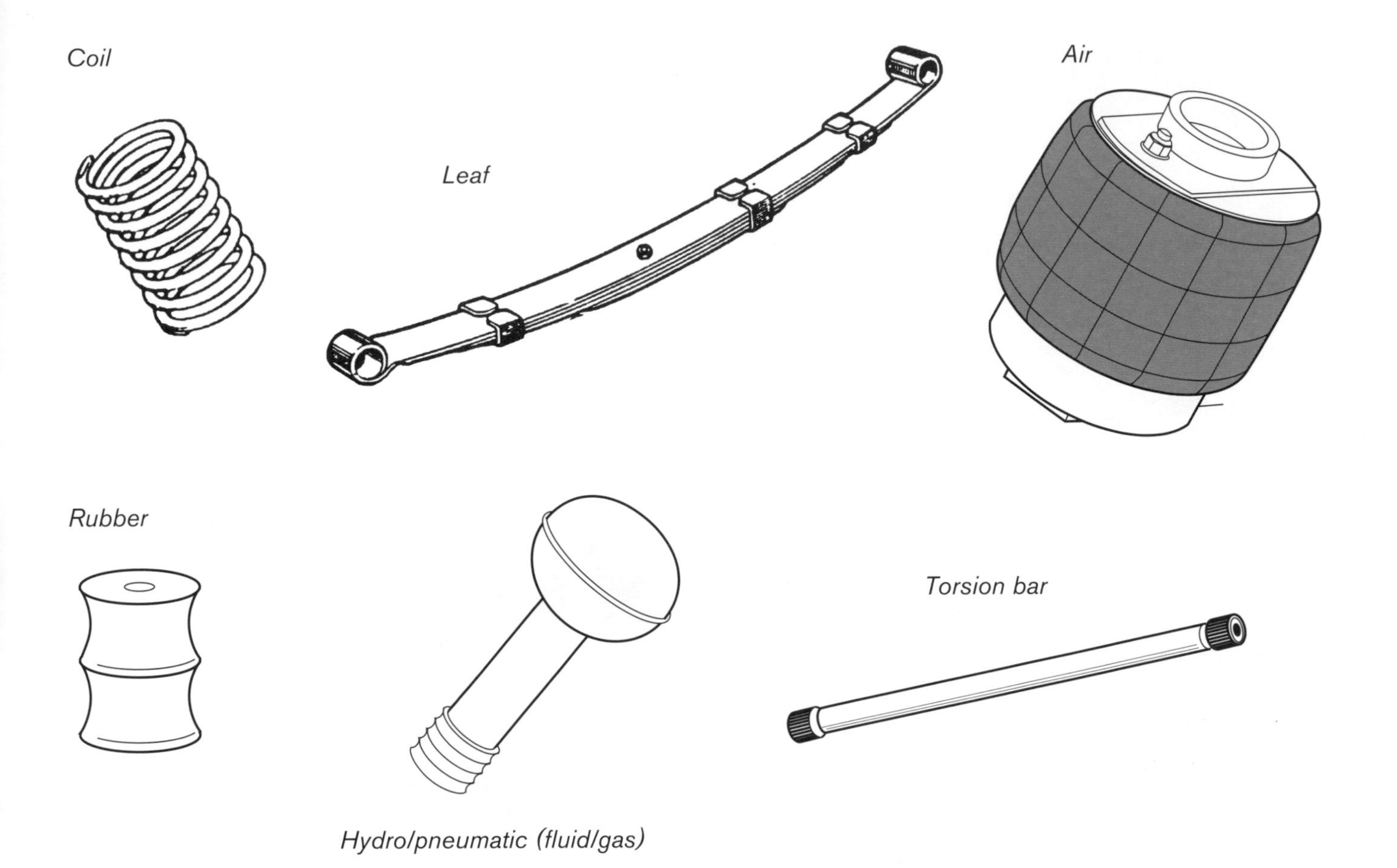

ACTIVITY

1 Name some current makes and models of cars, vans or HGVs that use the spring systems listed below. You can get this information from work or from magazines, or books in the library.

Spring system	**Make/model**
Leaf	
Coil	
Torsion bar	
Hydro/pneumatic	
Rubber	
Air	

2 Every vehicle body moves on its suspension under certain conditions. For example, it rolls during cornering. Other than road shocks, give two other conditions when the body would move.

a ....

b ....

ACTIVITY

Match the diagrams below with the names of the suspensions systems. Give examples of current vehicles that use them.

- MacPherson strut
- Multi-leaf
- Torsion bar
- Fluid/gas

	System	Vehicle
A		
B		
C		
D		

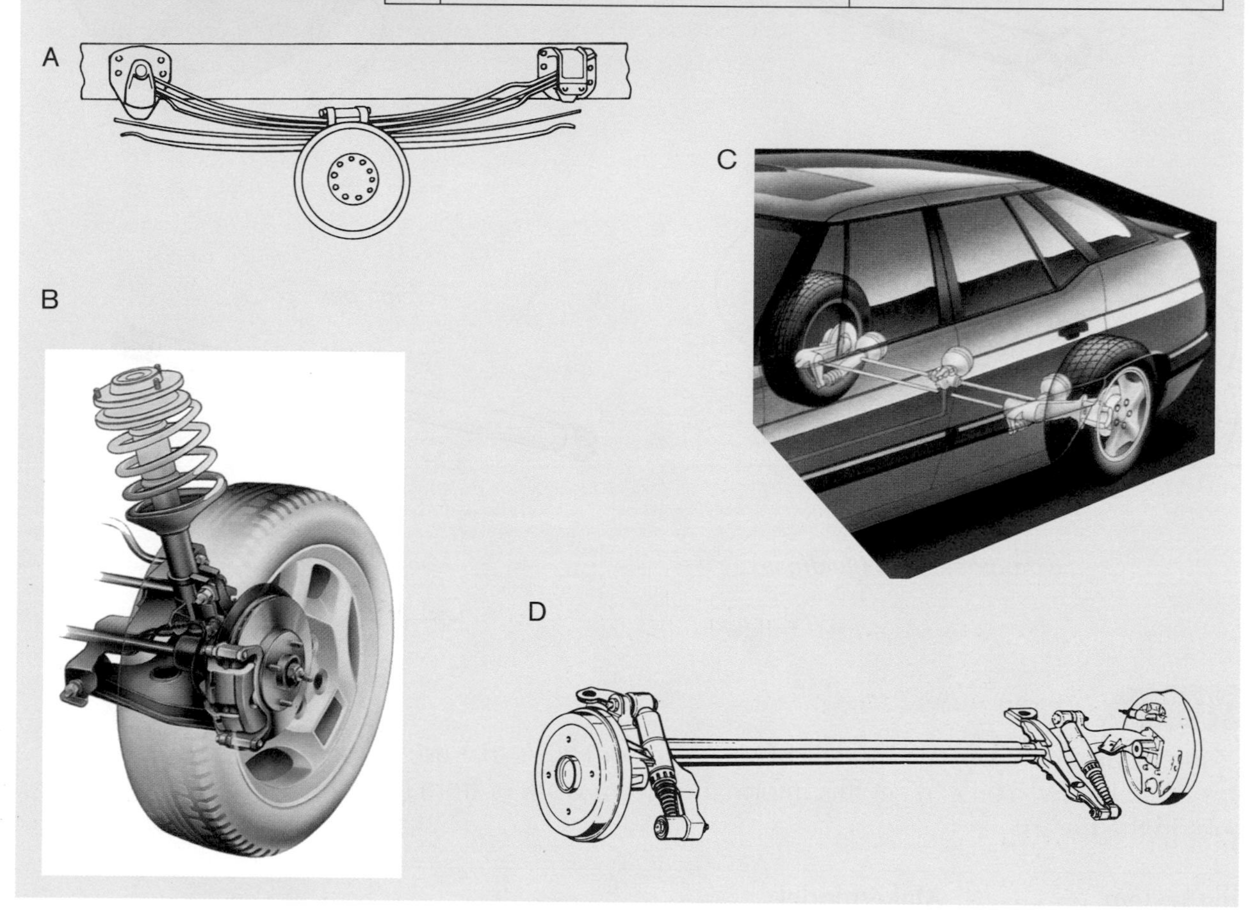

Shock absorbers (or dampers)

When a spring is pressed down and then released, it carries on flexing a few times before settling in its normal shape. So after going over a bump, the springs in a car would also bounce up and down. Bouncing would make the car difficult to control. It would be uncomfortable for people in the car, and things being carried might get broken.

To prevent this, a **hydraulic shock absorber** is placed close to the spring, between the axle and chassis. This stops the continued bouncing. The car can ride over a bump and stay stable.

There are two types of shock absorber: **strut** and **telescopic**.

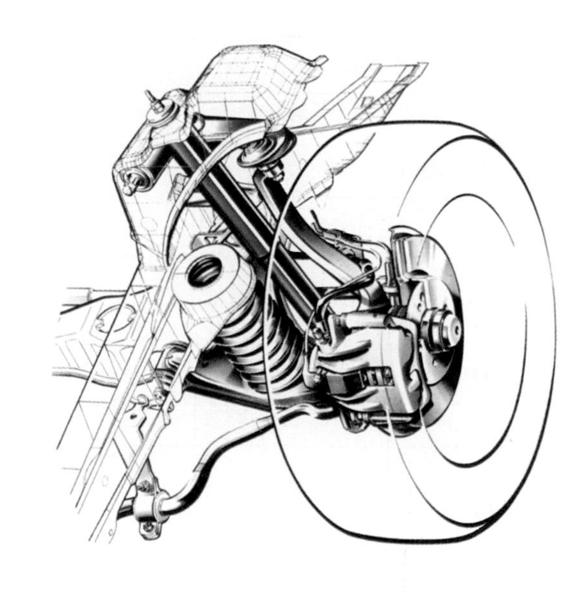

QUESTION

Name the types of shock absorber shown below.

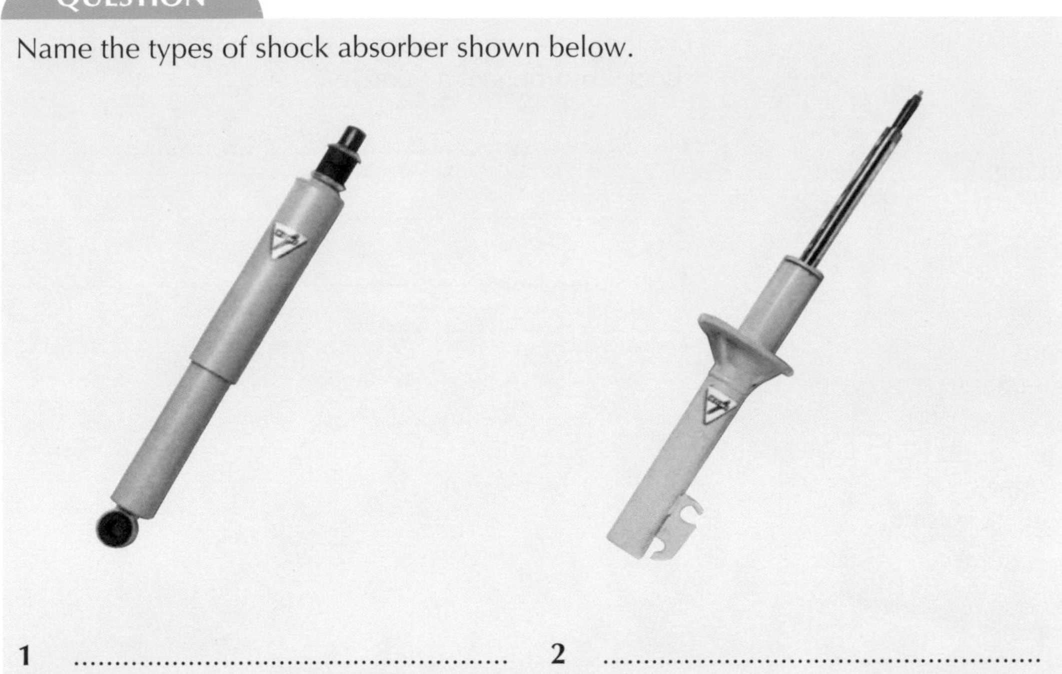

1 .... 2 ....

ACTIVITY

Hold a working telescopic shock absorber at each end. Try to operate it quickly. Is it easy to do so?

☐ yes ☐ no

It is impossible to operate a shock absorber quickly. You are trying to force a piston to move in a cylinder full of hydraulic fluid. Small holes in the piston (valves) let the fluid flow from one side to the other as the piston moves along the cylinder.

The shock absorber has a similar restricting effect on road spring movement.

A good suspension system still keeps all four tyres in good contact with the road surface, at all times. This is important for safe cornering, braking and accelerating.

piston rod

fluid

piston with bump and rebound valves

floating or free piston

gas

Faults and testing

One common fault with shock absorbers is **fluid leakage**. Without hydraulic fluid the shock absorber no longer works properly.

ACTIVITY

You can carry out a simple **bounce test** on a vehicle, to check whether the shock absorbers work well.

Start with one corner. Push down on the car body and let go. Count the number of bounces before the car comes to rest. Repeat for the other three corners.

There should not be more than 1.5 bounces.

HEALTH & SAFETY

When you are raising and supporting a vehicle, be sure to do this safely and correctly.

Jack Position the jack and axle stands at the correct place under the chassis. Chock the road wheels.

Vehicle lift Position the vehicle correctly and chock the wheels. Beware of obstructions when raising and lowering the lift.

Spring compressor Use an approved tool. Make sure that it remains properly clamped on the spring as you remove the spring.

Always keep your work area tidy and uncluttered.

Worn shock absorbers can have other effects:

- abnormal tyre wear
- vibration
- judder when braking
- noisy operation (knocking).

TIP
When you check shock absorbers, look at other components too – the chassis area, the springs, the suspension linkage, and so on. It is good if you can report any other problems. Don't just look at one thing – always watch out for other problems.

ACTIVITY

List the faults you might see when looking at the shock absorbers of a vehicle.

1 ....

2 ....

3 ....

4 ....

Shock absorbers can be tested using special equipment such as the Sachs **shock-tester** shown. This gives a print-out of the test results.

Types of shock absorber

Here are some other types of shock absorber:

- uprated
- mono
- adjustable
- active.

You can find out more about these from manufacturers' brochures (at work), car advertising booklets, and motor magazines.

ACTIVITY

Make short notes to explain:

1 Why *uprated* shock absorbers are used.
2 The difference between *mono-tube* and *twin tube* shock absorbers.
3 The advantages of *adjustable* shock absorbers.
4 The benefits of *active* suspension.

Keep these notes in your folder.

The **suspension kit** shown below (Koni) can be fitted instead of the original equipment. This gives a lower 'ride height' and a firmer ride, for sporty driving.

QUESTIONS

1 One effect of worn shock absorbers is:

a Vehicle leans to one side?
b Stiff steering?
c Front end vibration?

....

2 What is the function of the tool shown opposite?

....

....

3 The arrow on the drawing below shows the suspension part that connects the offside and the nearside suspensions. What is this part?

....

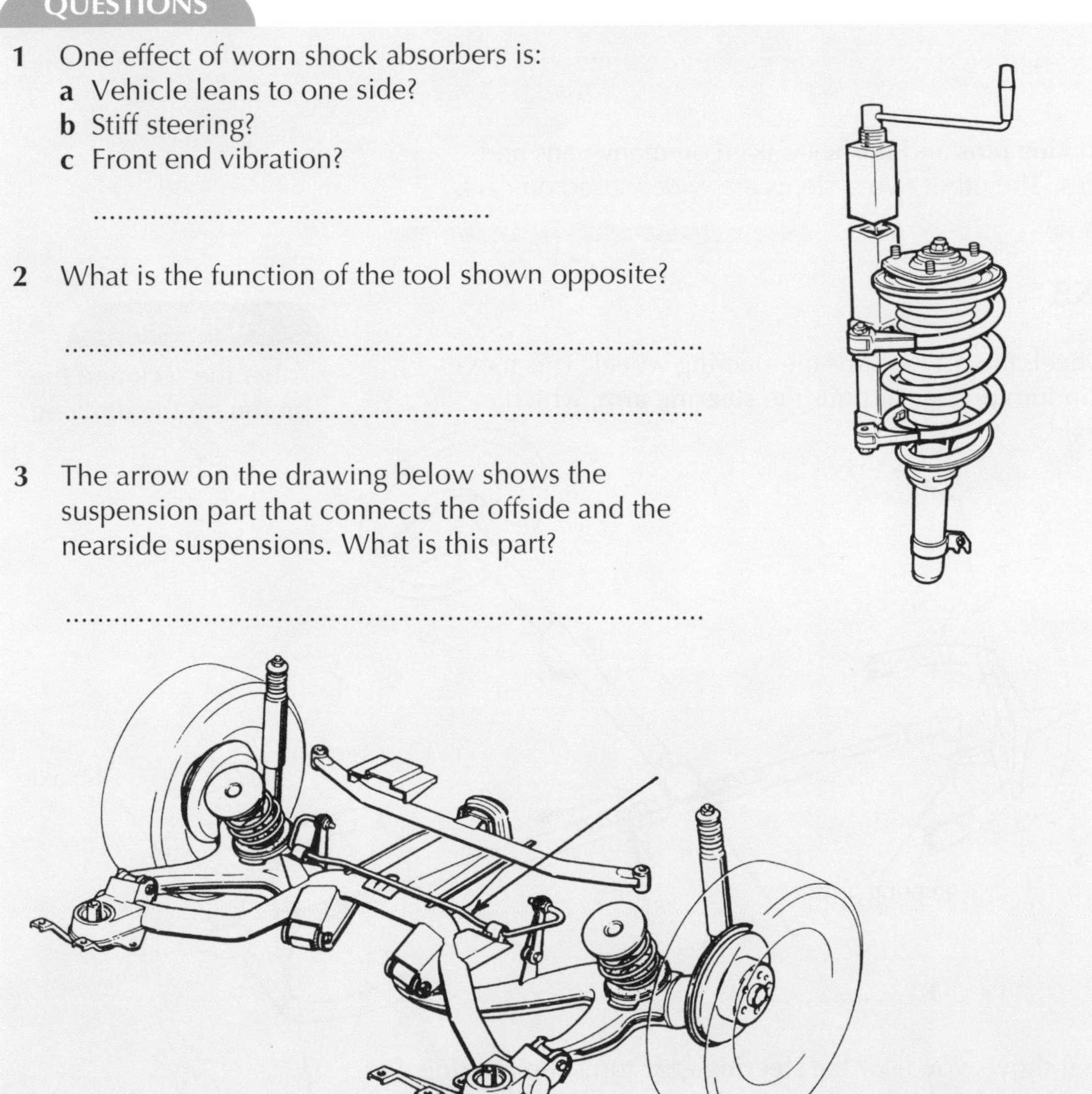

Steering

In steering a car, the front wheels have to turn to one side or the other – away from the normal straight-ahead position. To make this turn possible, the front wheel and hub are fitted on a **stub axle**. When you turn the steering wheel, the stub axle swivels.

QUESTION

Name the three types of steering arrangements below.

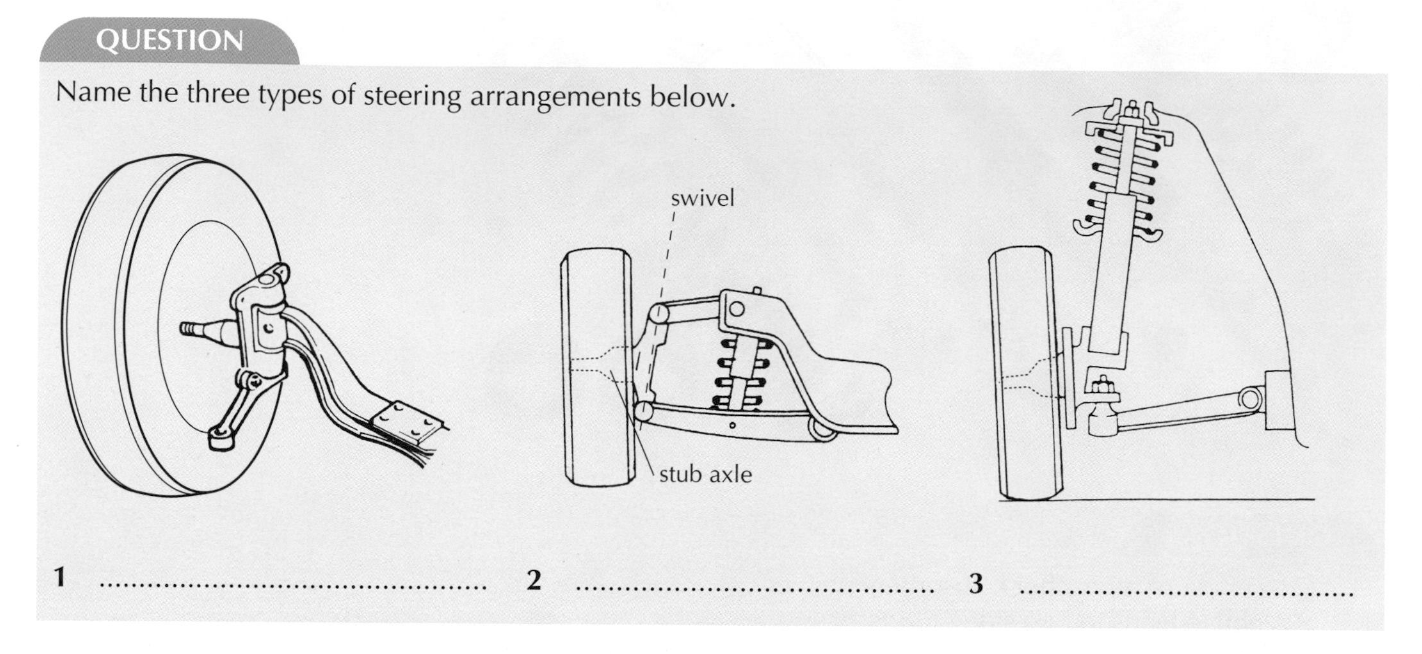

1 .... 2 .... 3 ....

The **beam axle** with **king pins** and **bushes** is used on many vans and heavy goods vehicles. The other two systems are widely used on cars.

How it works

ACTIVITY

Label the rack and the pinion on the drawing.

To steer the road wheel, the driver turns the steering wheel. This moves the **track rod**. This in turn pushes or pulls the **steering arm**, which is joined to the stub axle.

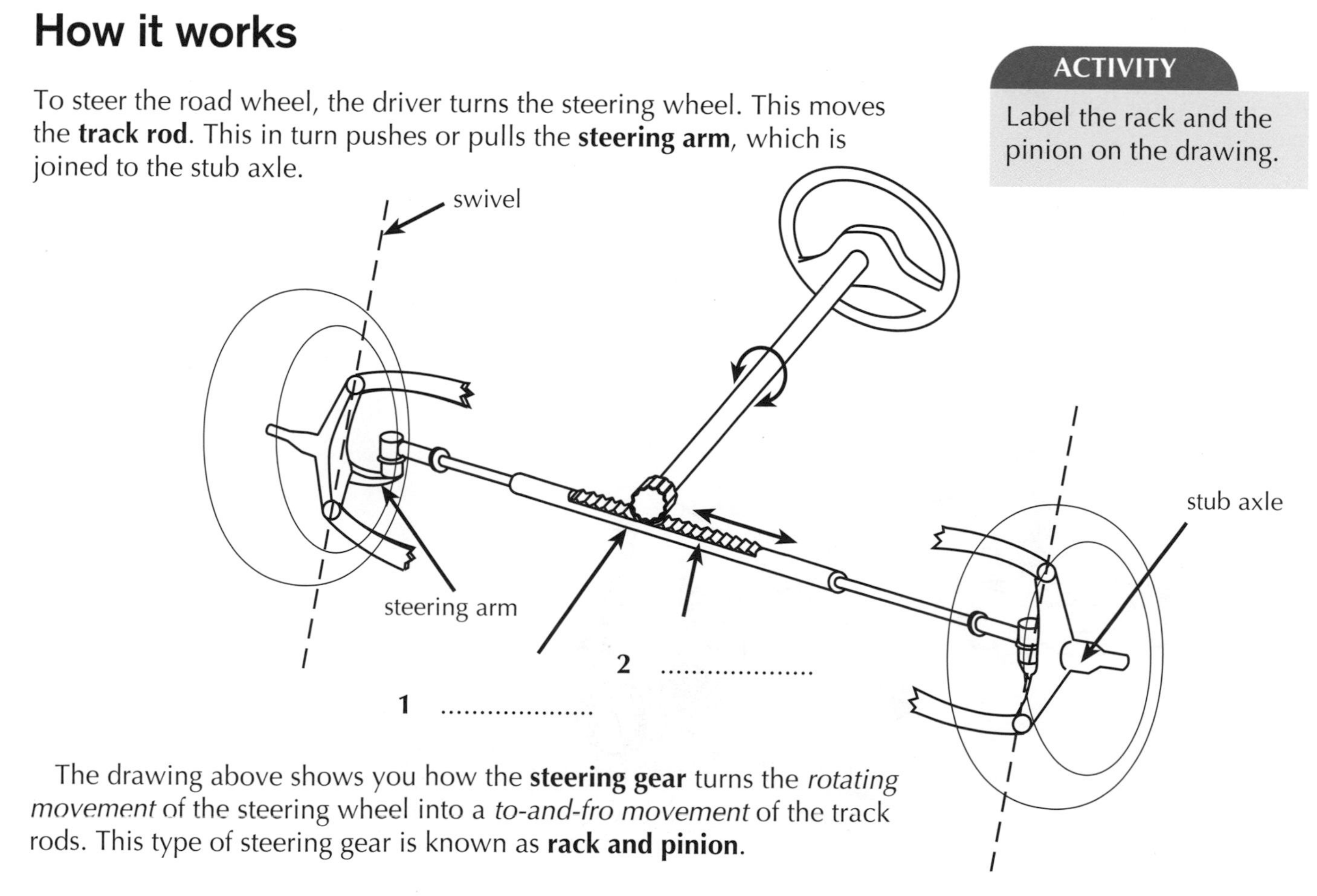

The drawing above shows you how the **steering gear** turns the *rotating movement* of the steering wheel into a *to-and-fro movement* of the track rods. This type of steering gear is known as **rack and pinion**.

This drawing shows the rack and pinion steering gear for a car. The rack slides inside a tube (the housing) and the pinion is mounted in a casing (another housing). All housings are filled with lubricating oil. The **bellows** on the rubber boots allow for track rod movement.

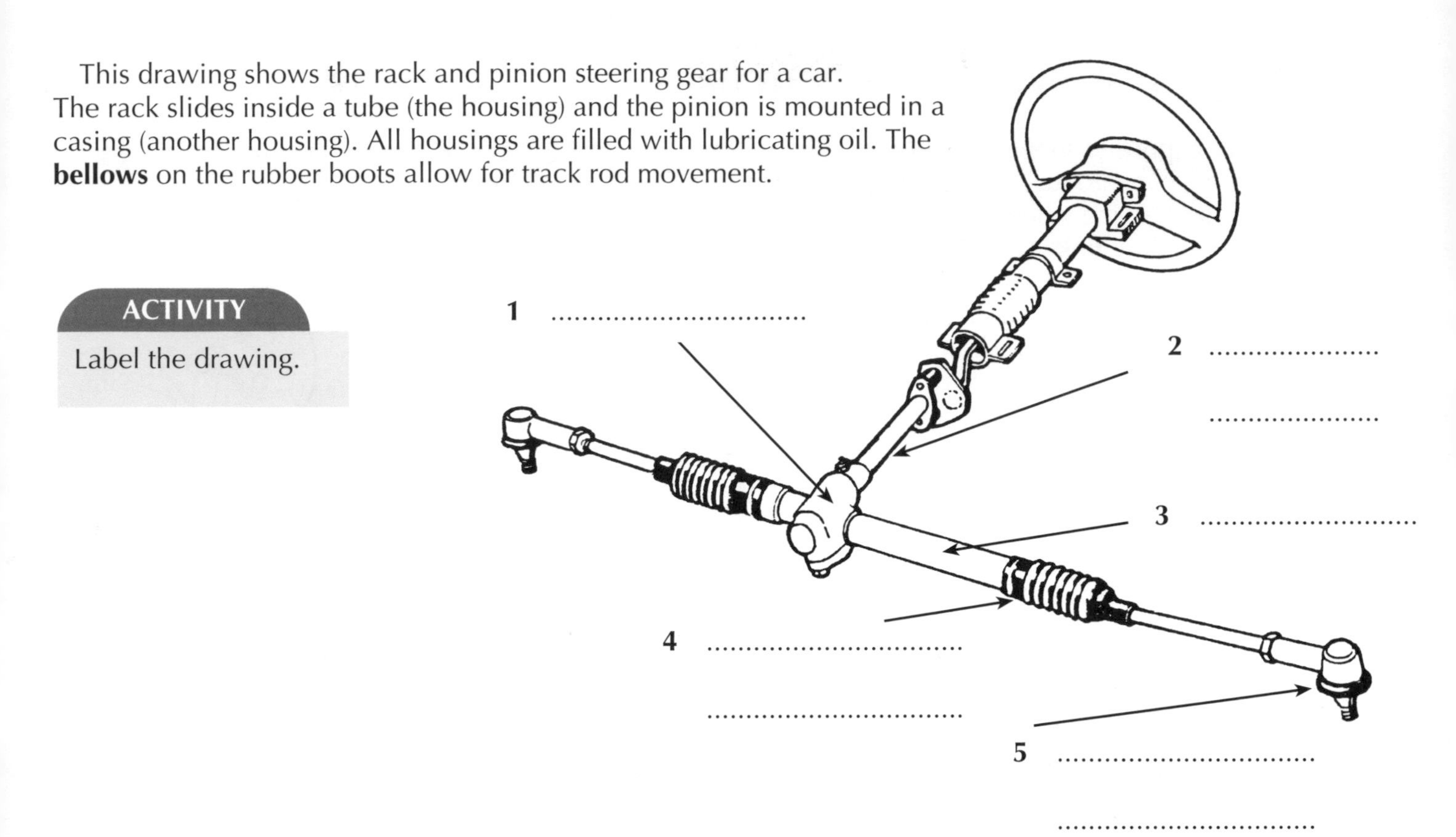

ACTIVITY

Label the drawing.

The drawing below shows in detail the parts of a steering rack that you might be likely to replace.

As you can see, the track rod end must be removed before you can remove and replace the rubber boot. The inner and outer linkage ball (joints 1 and 2) allow steering and suspension movement. The rubber boot shields the rack from dirt and dust as it moves in and out of the rack housing.

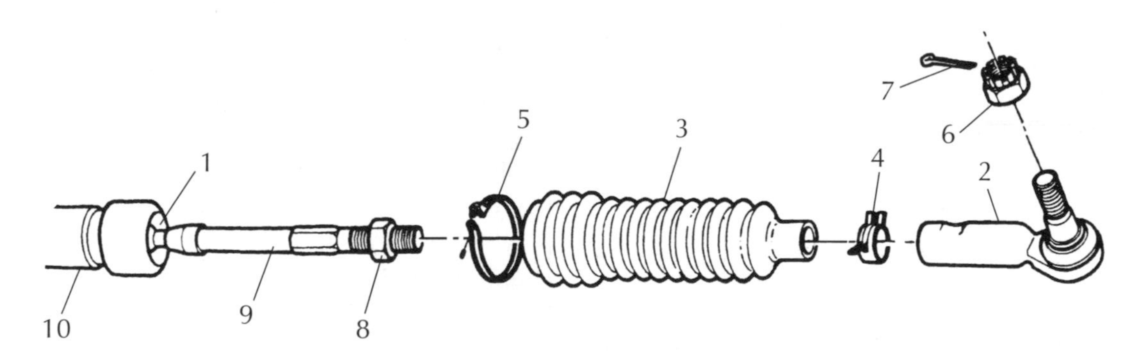

QUESTIONS

1 What other job does the rubber boot do?

....

2 Name the numbered parts.

1 .... 2 .... 3 .... 4 .... 5 ....

6 .... 7 .... 8 .... 9 .... 10 ....

Steering box systems (beam axle)

The diagram opposite shows the steering system for a typical goods vehicle.

Turning the steering wheel swings the **drop arm**. The **drag link** is joined to the stub axle: as the drop arm moves to and fro, so does the drag link. This steers the stub axle on one side. The movement is transferred to the *other* stub axle by the track rod.

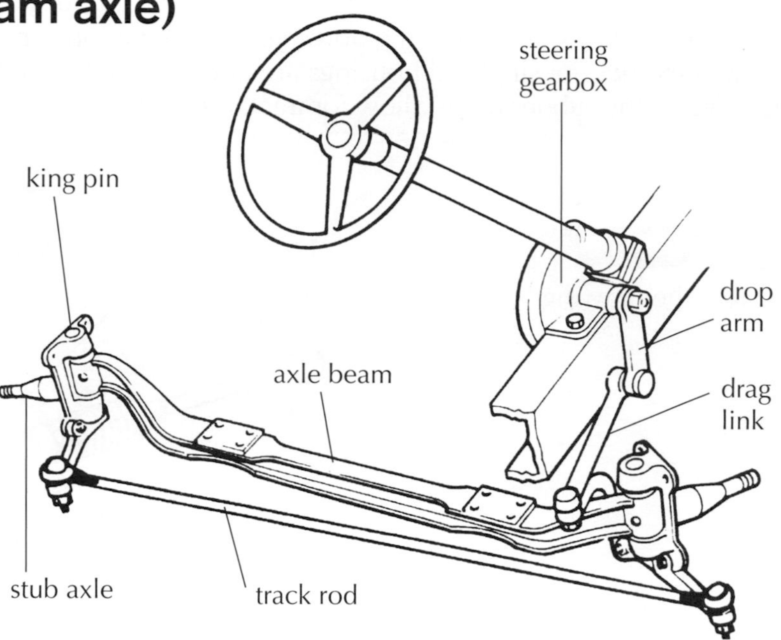

QUESTION

How many linkage ball joints are there in this system?

....

Three-piece track rod (IFS)

With **independent front suspension (IFS)** a steering box may be used. The track rod then has three parts. The **steering idler** transfers the drop arm movement to the other track rod.

ACTIVITY

Label the drawing.

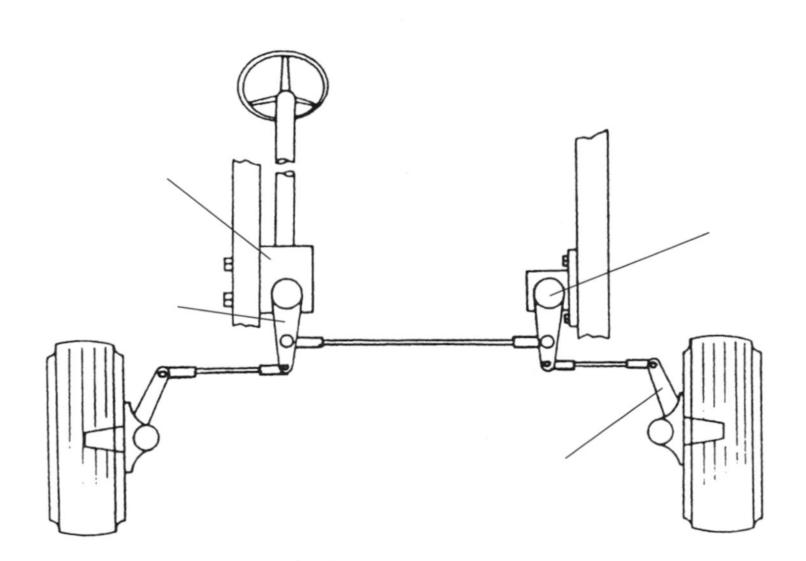

WORKSHOP

Examine vehicles with different steering systems. With the weight off and the road wheels free, watch what happens when the steering wheel is turned. Notice the stub axle swivel and linkage movement.

Steering service and repair

The most common faults in steering systems are:

- worn linkage ball joints (e.g. track rod ends)
- stub axle swivel wear (kingpins and bushes)
- worn suspension ball joints
- worn steering gear (rack and pinion or box)
- oil leakage.

To check the linkage ball joints, get someone to 'rock' the steering wheel with the weight on the road wheels. Look for up-and-down movement at **A** (see diagram).

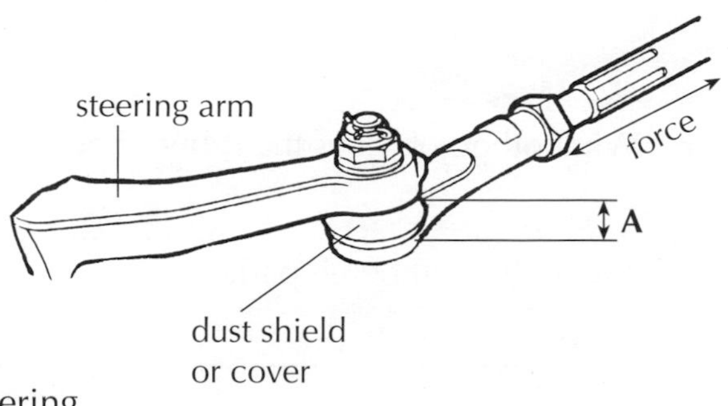

QUESTION

What else would you check on the ball joints?

....

QUESTION

This tool can be used when removing ball joints. What is its name?

....

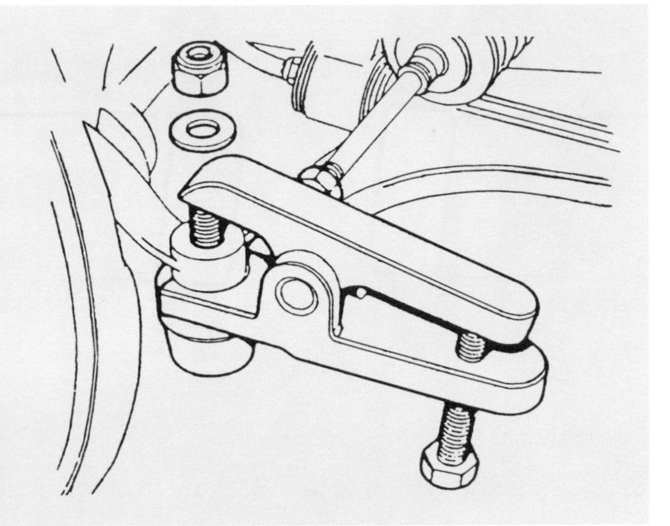

TIP
After replacing ball joints, always check the wheel alignment.

TIP
When you replace a track rod end, screw it on until you reach the same position as the old one. The lock nut is the guide. If you do this the wheel alignment will need little or no adjustment.

Wheel alignment (tracking)

The wheels on either side of a vehicle must be in line with each other when *rolling straight ahead.* That is, they must be parallel with each other as shown opposite.

The **wheel alignment** on the steered wheels is adjusted and set to the manufacturer's specification. This is a **static setting** – you make it when the vehicle is stationary.

The setting varies from vehicle to vehicle. It depends on the vehicle's design, and on the type of steering, the suspension, the tyres, the transmission and so on.

For example, a typical **toe-in** setting is 1 mm. As the vehicle is driven along the road the wheels would tend to roll outwards very slightly – about 1 mm. They would then be in line with each other. This would save undue wear on the tyre treads.

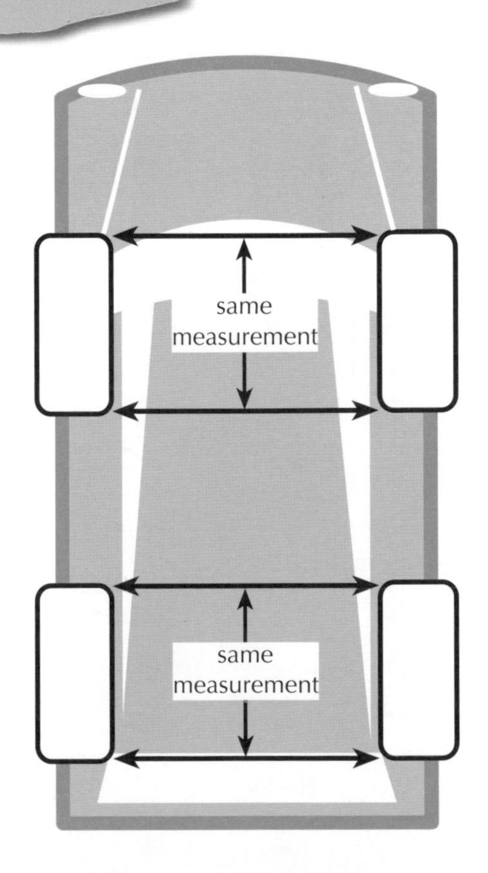

QUESTION

On the drawings below, say which is *toe-in and which is toe-out.*

Front

A

A = (B − 1 mm)

B

A

A = (B + 1 mm)

B

1 .... **2** ....

Checking and adjusting wheel alignment

A popular, simple, wheel alignment gauge is shown below.

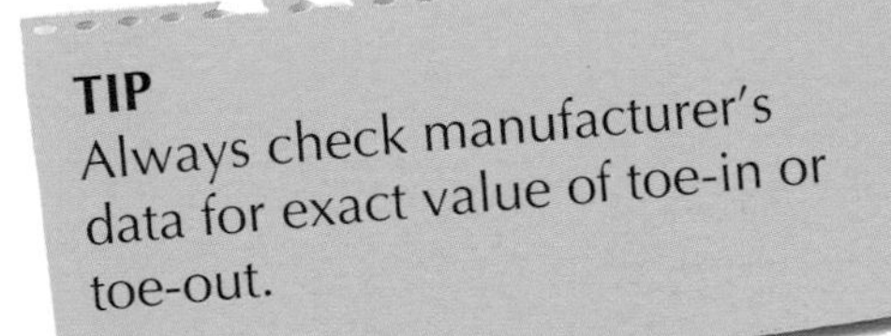
TIP
Always check manufacturer's data for exact value of toe-in or toe-out.

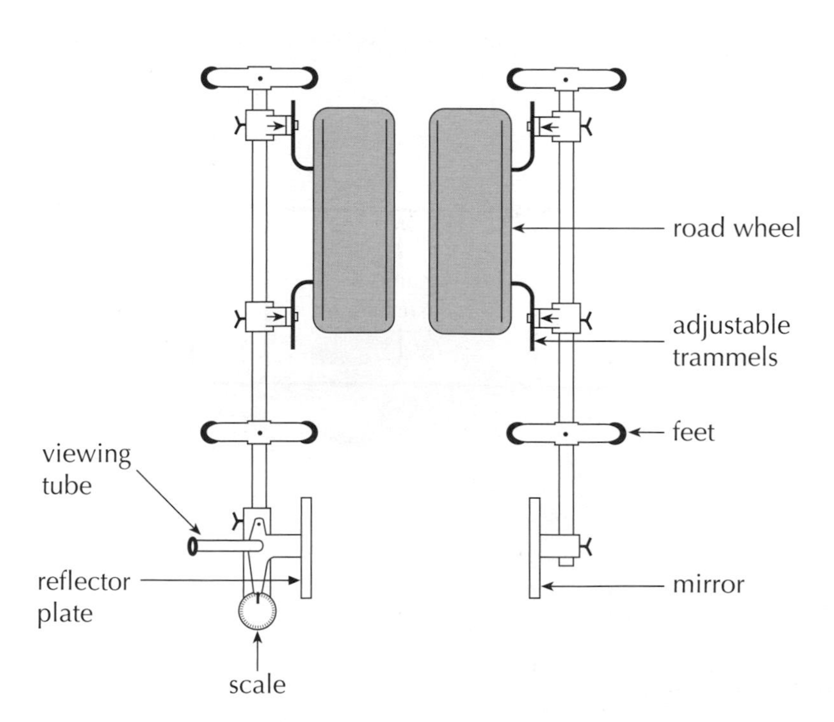

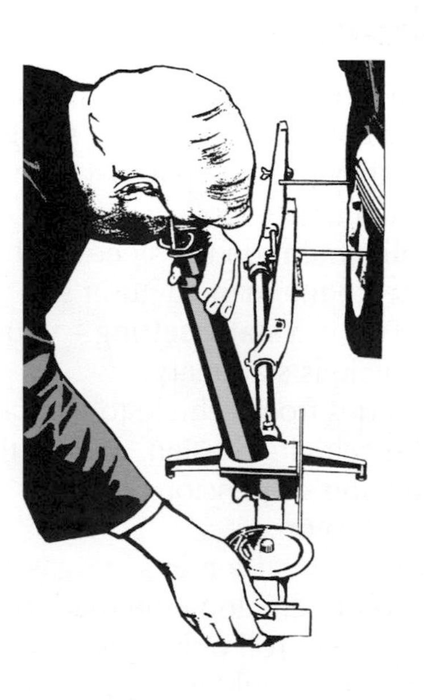

Different types of optical wheel alignment equipment

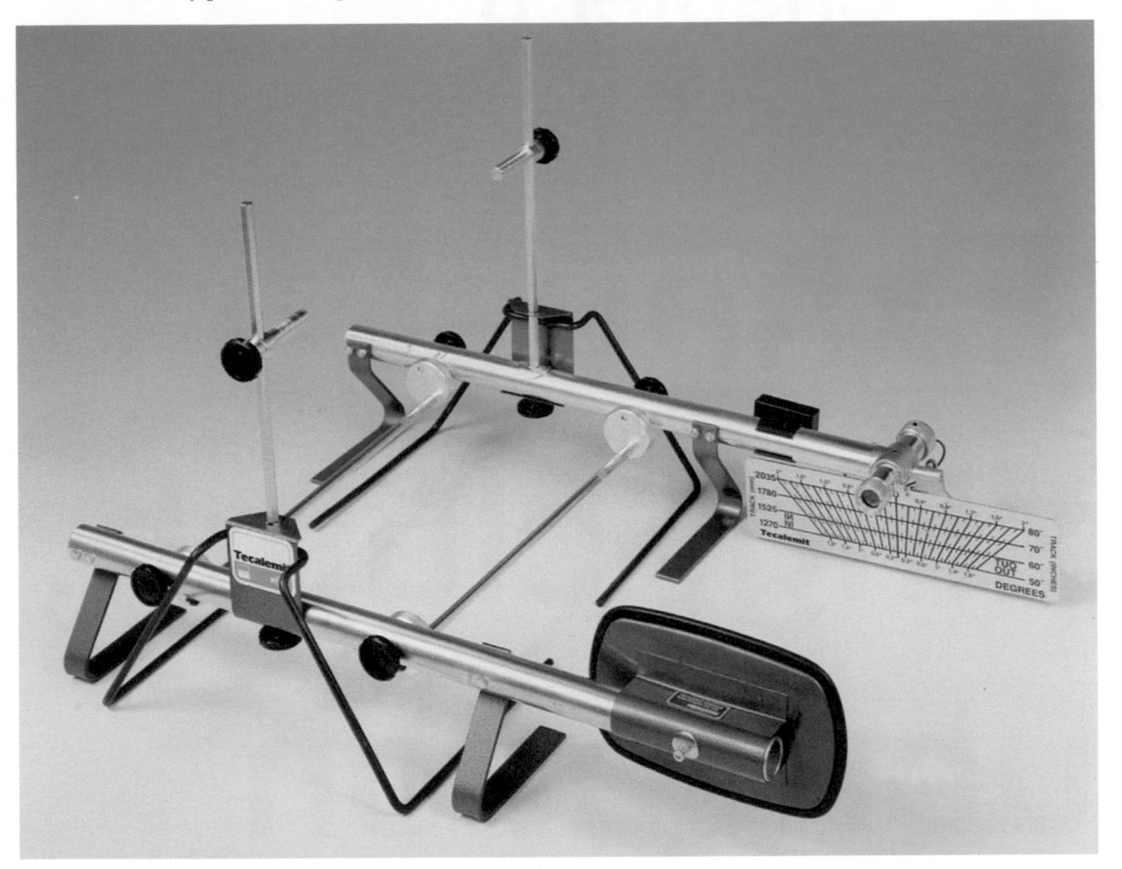

Set-up for calibration

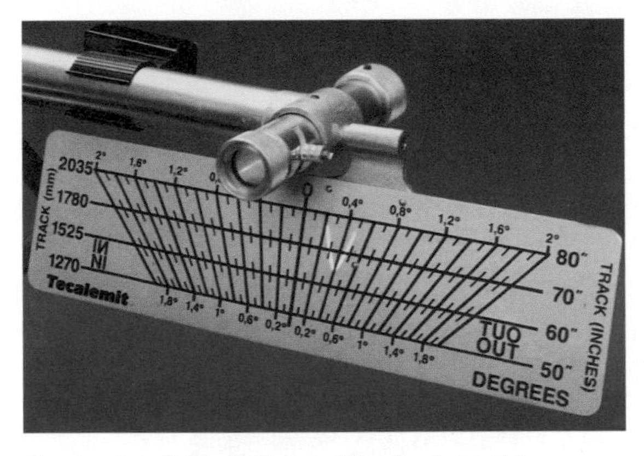

Scale detail from above. The bulb in this instrument has a special filament which throws a 'V' pattern on the scale below the projector. This helps accurate positioning. You can adjust the mirror so that the pattern is projected onto the right part of the scale.

TIP
Before checking the wheel alignment or adjusting it, check the tyre pressures.

Also check the condition of all ball joints (linkage and suspension). A worn ball joint or a bent track rod will directly affect the wheel alignment.

Chapter 8

Exhaust, cooling, clutch, lubrication

In this chapter you will learn about:

- **exhaust systems** – maintenance of pipes, silencers and joints
- **cooling systems** – general maintenance and antifreeze
- **clutch operation** – clutch control and adjustment
- **lubrication** – the many different lubricants used, and the various filters

As the engine and transmission are used, parts become worn. Some parts need to be replaced after a certain time or a certain mileage.

The exhaust system carries the exhaust gases away from the vehicle, so that they do not poison the driver and passengers. It also reduces the noise.

This chapter helps you to understand these systems, and explains how to check and service them.

Exhaust systems

The **exhaust system** quietens or silences the noise made by exhaust gas leaving the engine.

The main parts of an exhaust system

front down pipe

tail pipe assembly

flexible pipe or joint

catalytic converter

centre section including two silencer boxes of different designs

ACTIVITY

Name the parts of the exhaust system.

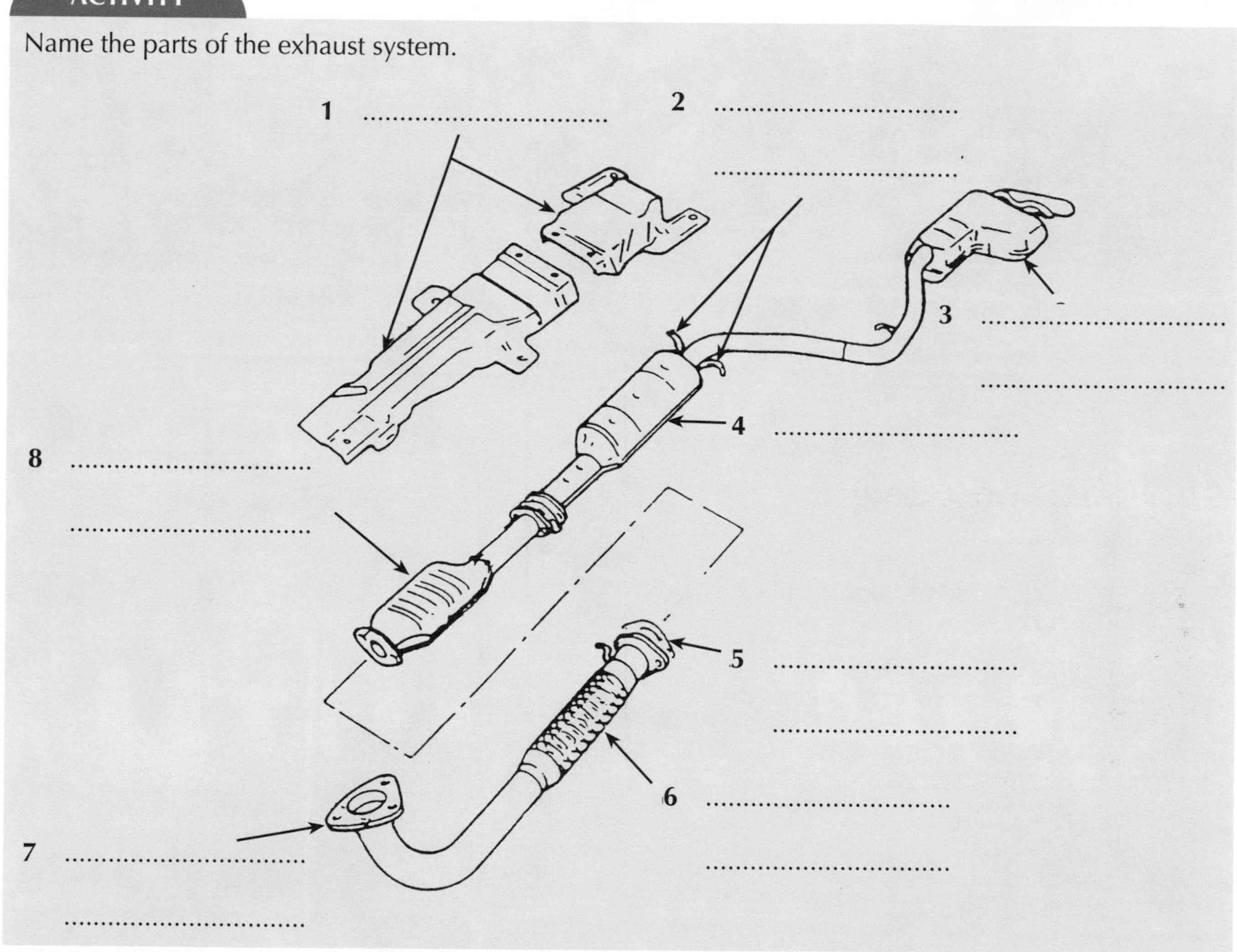

The MOT test

For exhaust systems, the MOT includes these checks:

- **Condition** – check for signs of corrosion and security of mountings.
- **Leaks** – check for leaks from joints, corroded pipes or boxes.
- **Noise** – check for noise levels greater than expected for that vehicle.
- **Exhaust emissions** – check for carbon monoxide and hydrocarbons: emissions must be below certain maximum values.

ACTIVITY

1. Using MOT tables, find out the percentage limits for a vehicle's exhaust emissions.
2. Make sure that you understand the meaning of the various terms.
3. If possible, get an MOT exhaust gas emission read-out sheet. Add this to your portfolio.

Environmental protection

Modern exhaust systems remove harmful pollutants from the engine. This is done by the **catalytic converter**.

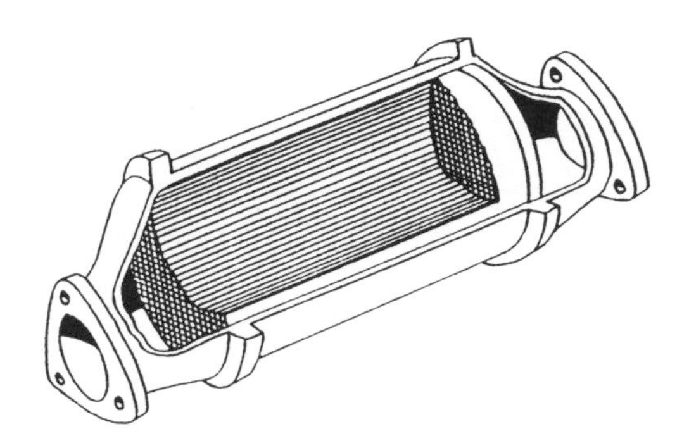

HEALTH & SAFETY

Hot exhaust systems can cause serious burns.

Allow exhaust systems to cool before inspecting. Note that the catalytic converter gets extremely hot, and it takes a long time to cool.

Inspection and testing

WORKSHOP

Place a vehicle on a ramp. Examine the exhaust system for faults.

Look for:

- signs of excessive corrosion or defects – these might allow exhaust fumes to get into the vehicle
- loose connections at the joints
- loose mountings
- broken or missing exhaust supports
- bent or damaged system
- contact between the exhaust system and the vehicle body.

Is it necessary to replace all or part of the system?

Before fitting anything, make sure that you are using the correct parts.

TIP

- When *removing* the system, work from the back to the front of the vehicle. When *replacing* the system, start at the front.
- When replacing the system, use new rubber mountings, gaskets and clamps.
- When reassembling, apply anti-seize compound to all nut and bolt type fasteners.

ACTIVITY

Name the types of clamps and supports indicated.

fibre

rubber

1 ....

2 ....

3 ....

4 ....

HEALTH & SAFETY

When running an engine in the workshop, always use an exhaust gas extraction system.

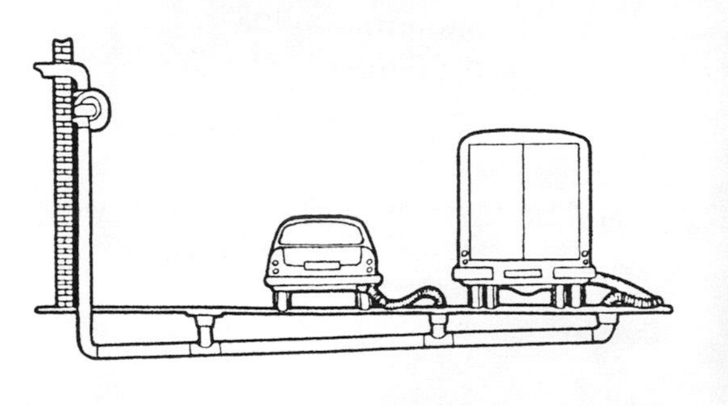

An underground exhaust extraction system

WARNING

At quick-fit repair workshops it is usual to cut corroded exhaust systems using oxyacetylene welding equipment.

This method takes skill, and understanding of fire safety. It needs experience.

Do not attempt such work without supervision.

Exhaust pollutants

Exhaust fumes from vehicles contain several pollutants which can damage health. The percentages show how much of each comes from exhaust fumes.

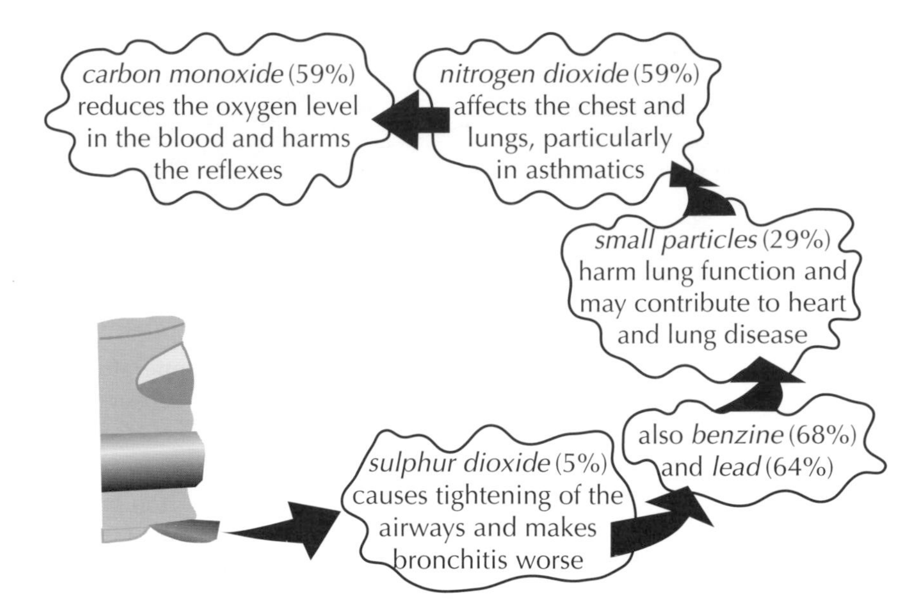

Engine cooling

To work, an engine must burn fuel. This generates a great deal of **heat**. The cylinders could get so hot that the metal might melt. To prevent this, the engine is cooled.

There are two main types of cooling systems. **Air cooling** is found mainly on motor cycles and the old style rear-engined Volkswagen Beetle. **Liquid cooling** is used for cars and commercial vehicles.

Heat will flow from a hotter place to a cooler place. It is transferred in three ways:

- **Conduction** – the heat passes through solid materials, mainly metals.
- **Convection** – the heat is carried by moving liquid or gas.
- **Radiation** – the heat is given off into the air, from the surface of the object.

In a vehicle's **cooling system** all three kinds of heat transfer happen at the same time.

Liquid cooling systems

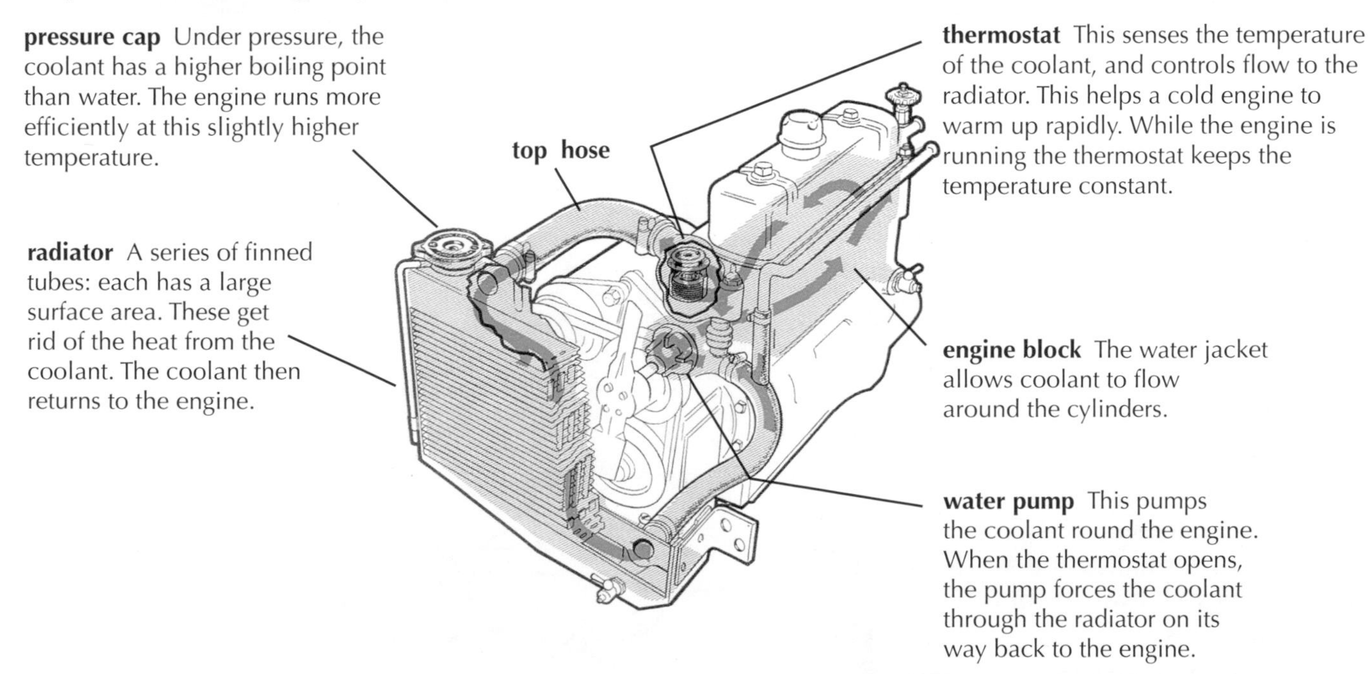

An engine mounted in line with the chassis (engine at the front; rear-wheel drive)

electric fan This draws cool air through the radiator. It is switched on when needed. Sometimes you will hear it still working when the engine has been stopped.

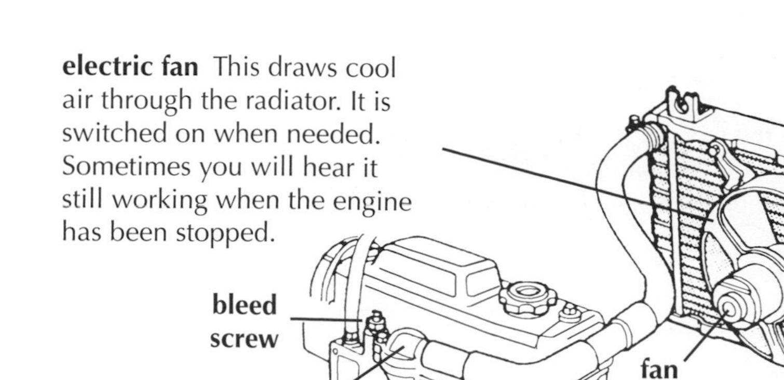

expansion tank This gives space into which the heated coolant can expand. Because of this no coolant is lost.

top hose, bottom hose Hoses connect the engine and the radiator. The radiator is fixed to the vehicle's body, and the engine vibrates when it is running, so the hoses need to be flexible.

A car engine, mounted sideways, or transversely (engine at the front; front-wheel drive)

HEALTH & SAFETY

Never put your fingers near the fan, even when the engine is stopped. The fan might start automatically.

Inspection and testing

WORKSHOP

1. Inspect a cooling system. Check that:
   - there are no leaks
   - the hoses are in good condition
   - fluid is at the right level.
2. Replace any faulty items.

HEALTH & SAFETY

When topping up a cooling system, **do not remove the pressure cap when the engine is hot**. The temperature may well be over 100 °C. Any steam given off will be hotter still.

Keep your hands and any loose clothing well away from rotating engine parts.

TIPS

Check hoses for burned or chafed areas. Always feel under the hoses. Squeeze the hose to find any cracks and breaks – these could cause leaks.

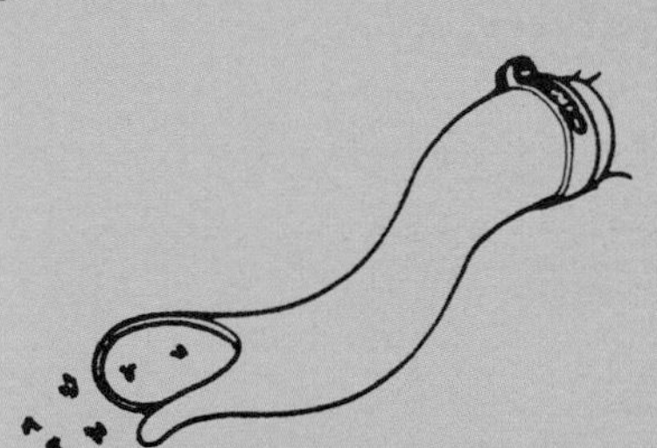

A *soft hose* means that there is decay or wear. Inside the hose, bits may break off. These might get into the cooling system and clog the radiator.

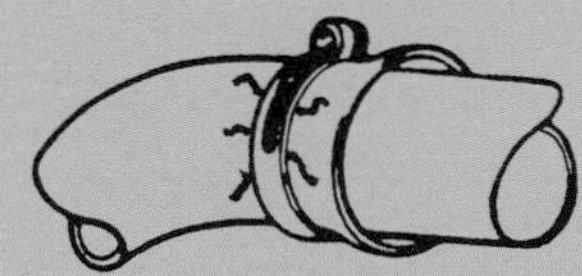

Hard hoses can crack and fail at any time.

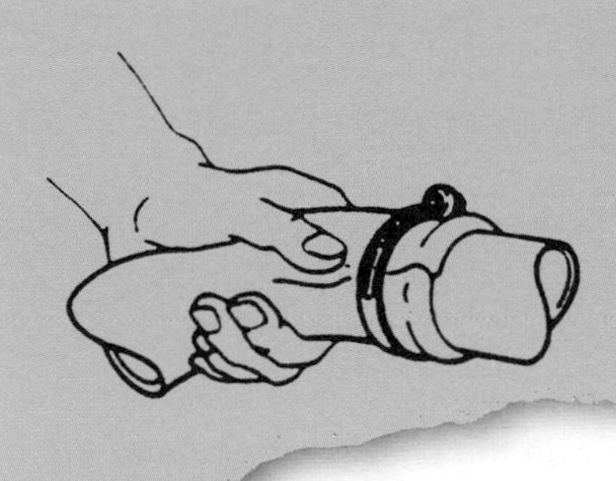

A *swollen hose* or oil-soaked ends are signs of danger. They mean oil contamination and possible failure.

WORKSHOP

1 Remove the pressure cap. Check that the valves are operating correctly. Look for any signs of weakness.

2 Pressure-test the cap for correct opening pressure and seal leakage.

Testing the radiator cap Connect the cap to the tester. Pressurise the tester. Check when the valve opens.

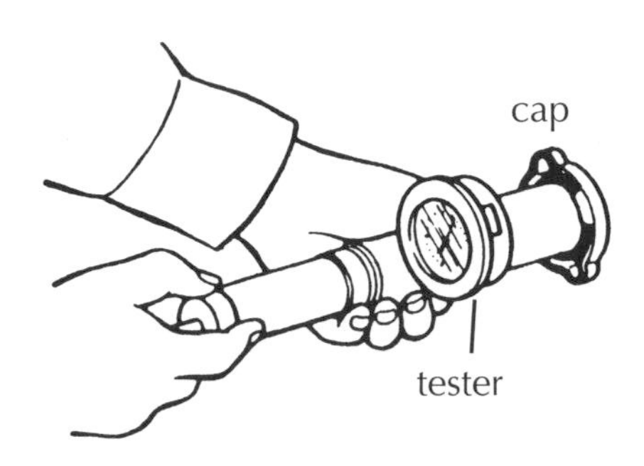

Cap relief pressure was:..

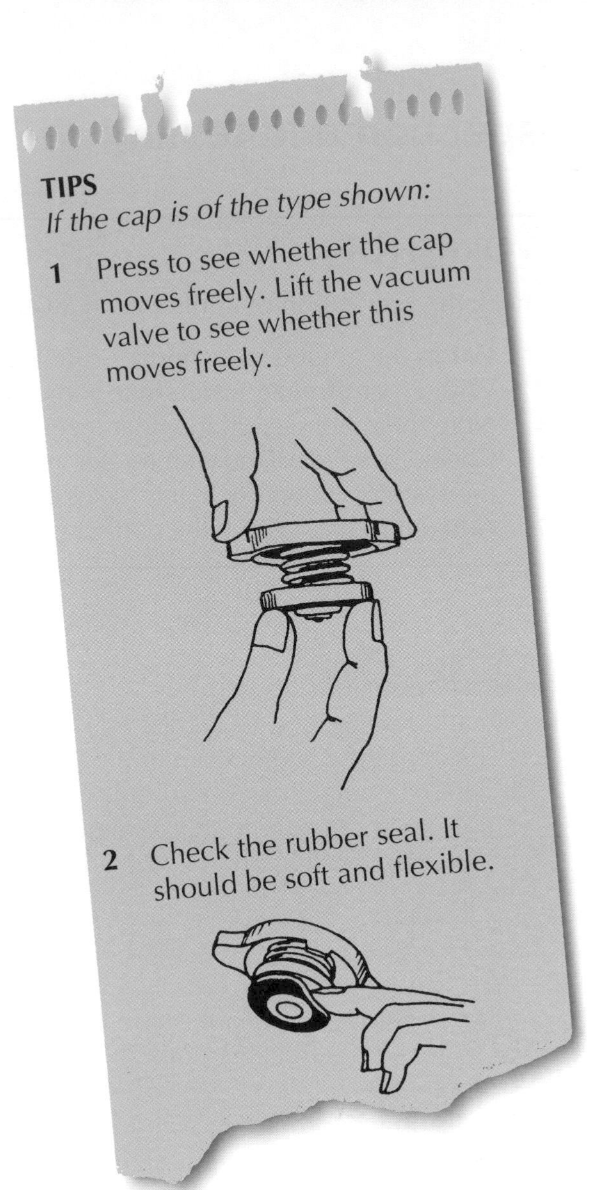

TIPS

If the cap is of the type shown:

1 Press to see whether the cap moves freely. Lift the vacuum valve to see whether this moves freely.

2 Check the rubber seal. It should be soft and flexible.

Effects of freezing

When water freezes, the volume gets bigger. This expansion applies great force. This can crack the engine cylinder block and leak. It could well mean a new engine!

Antifreeze

Look at the graph below. The freezing point of water can be changed by adding certain liquids.

Ethylene glycol makes the freezing point *lower* – the coolant is then less likely to freeze. Before freezing, it forms a mush.

Additives in the antifreeze help to stop corrosion of metal in the cooling system.

% ethylene glycol in coolant

0 10 20 30 40 50

temperature °C

−10 −20 −30 −40 −50

liquid

MUSH

solid

HEALTH & SAFETY

If antifreeze is spilt on paintwork, wash it off immediately. Antifreeze will strip paint.

Antifreeze can be toxic (poisonous).

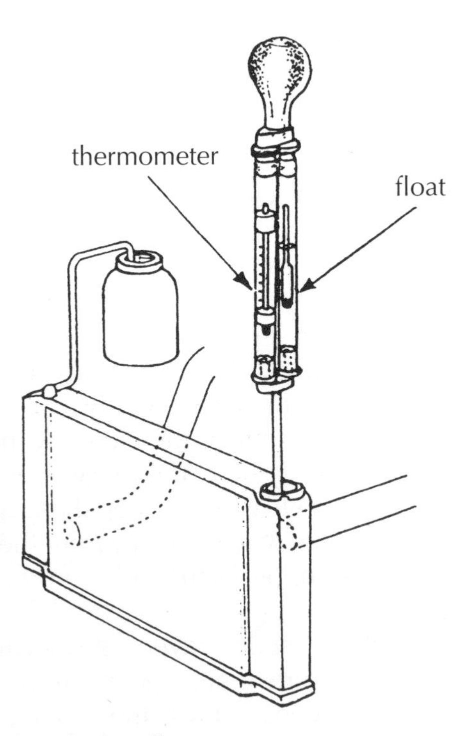

A typical antifreeze tester

Inspection and testing

WORKSHOP

Check the proportion of antifreeze in a vehicle's cooling system.

1 Warm the engine to its normal running temperature.
2 With an **antifreeze tester**, take some coolant from the system.
3 Note the readings, on the float level and on the thermometer.
4 Check these readings with a reference card. What percentage of antifreeze is there in the system? How much antifreeze do you need to **top up** to a safe percentage?
5 Add antifreeze to give the correct percentage.

QUESTIONS

1 What does the thermostat do?
a Prevent the coolant from boiling?
b Help the engine warm up quickly?
c Maintain pressure in the system?
d Raise the boiling point of the coolant?

....

2 Why is antifreeze added to the coolant?
a To prevent the coolant from freezing?
b To lower the freezing point of the coolant?
c To stop ice forming in the radiator?
d To stop the water pump seizing up?

....

3 Which part lets liquid-cooled systems operate at high temperatures?
a The thermostat?
b The pressure cap?
c The water pump?
d The temperature gauge?

....

4 What is a suitable running temperature for the coolant in a liquid-cooled engine?
a 48 °C?
b 88 °C?
c 118 °C?
d 148 °C?

....

The clutch

The **clutch** is part of the **transmission system**. It is a **coupling**: it connects the drive from the engine to the gearbox. The drawing opposite shows where the clutch is.

The driver controls the clutch with a **clutch pedal**. Pressing the pedal releases or disengages the clutch. Releasing the pedal engages the clutch. In this way the drive from the engine is connected to, or disconnected from, the gearbox.

To select first and reverse gears with the car stopped, and to change gears during driving, the driver needs to press the clutch pedal before moving the **gear lever**.

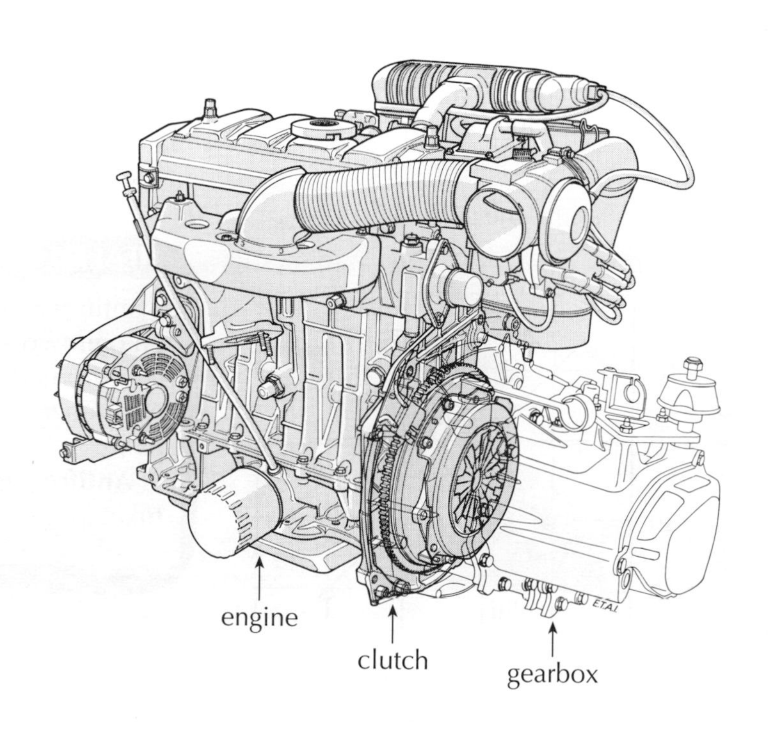

These very much simplified drawings show how the clutch works.

When it is transmitting drive (**engaged**), the **clutch plate** is clamped between the **flywheel** and the **pressure plate**.

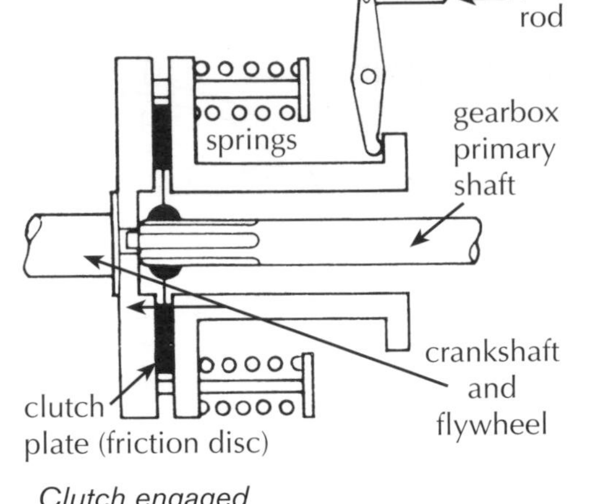

Clutch engaged

QUESTION

How is drive passed to the gearbox shaft?

....

....

To **release** the clutch plate, the pressure plate is withdrawn. The engine is still running, so the flywheel and the pressure plate are still turning. The clutch plate and the gearbox shaft are *not* turning.

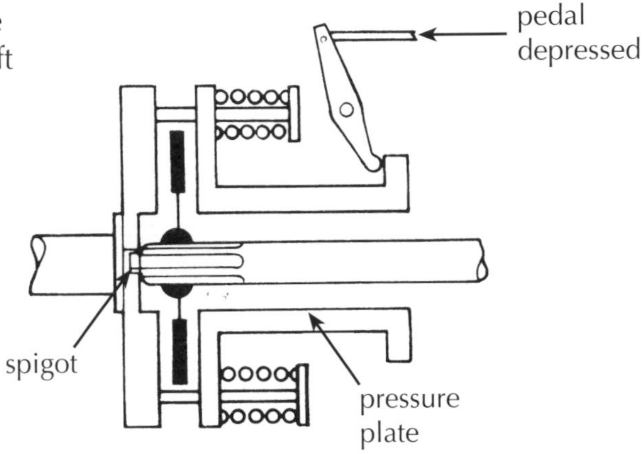

Clutch released

QUESTION

The clutch also allows the driver to start the car from rest gradually and smoothly. How?

....

....

ACTIVITY

The main parts of a clutch are, **A** the clutch plate; **B** the pressure plate; **C** the release bearing; and **D** the flywheel. Add letters to the drawing below to match these names.

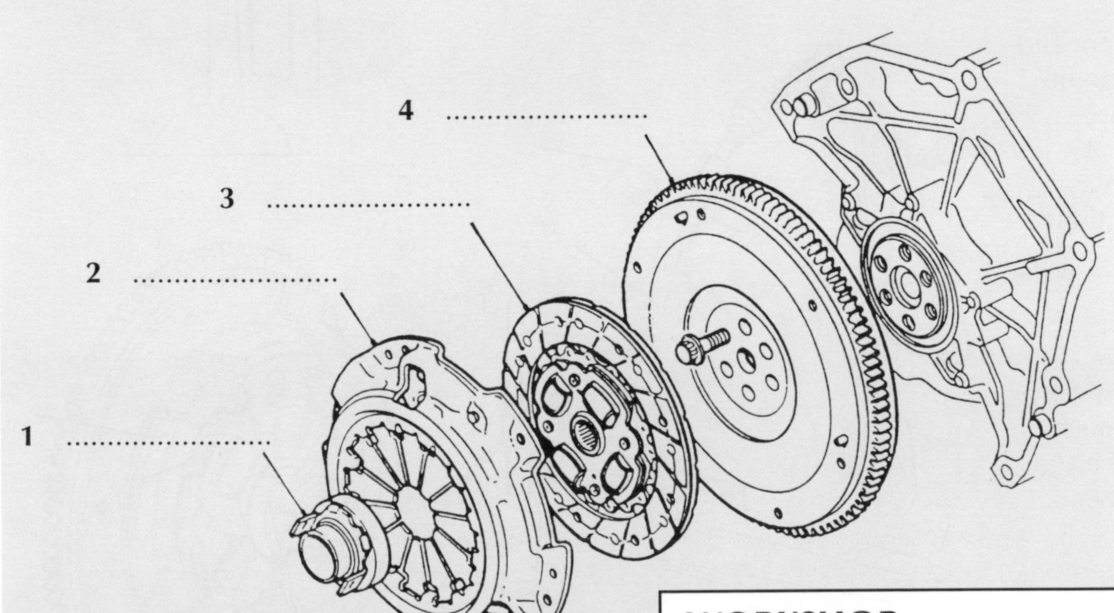

WORKSHOP

Look at a real clutch and examine the parts shown in the drawing above.

Notice the **friction linings** and the **splined hub**.

Clutch control

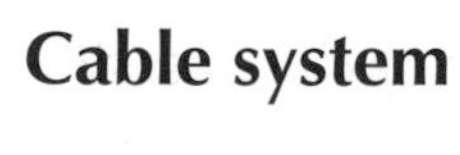

Cable system

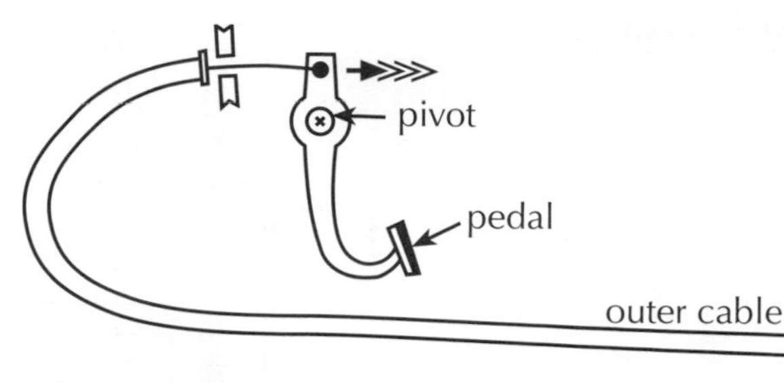

With a cable control system, the clutch pedal lever pulls the **withdrawal lever** and **release bearing**. The bearing is forced against the pressure plate diaphragm spring. This causes it to flex, releasing the clutch plate.

Hydraulic system

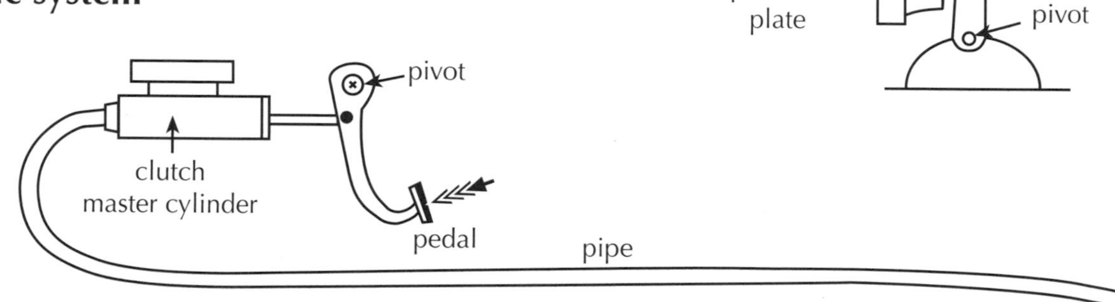

With a hydraulic clutch control system, when the pedal pumps fluid into the system this causes the **slave cylinder piston** to move. This moves a rod which operates the withdrawal lever.

Clutch adjustment

Both the cable system and hydraulic system can be adjusted. The release bearing must be kept in the right position as the clutch plate friction linings wear.

In most modern systems, adjustment is automatic. Most manual adjusters on cable systems are much like the adjuster for a bicycle brake cable.

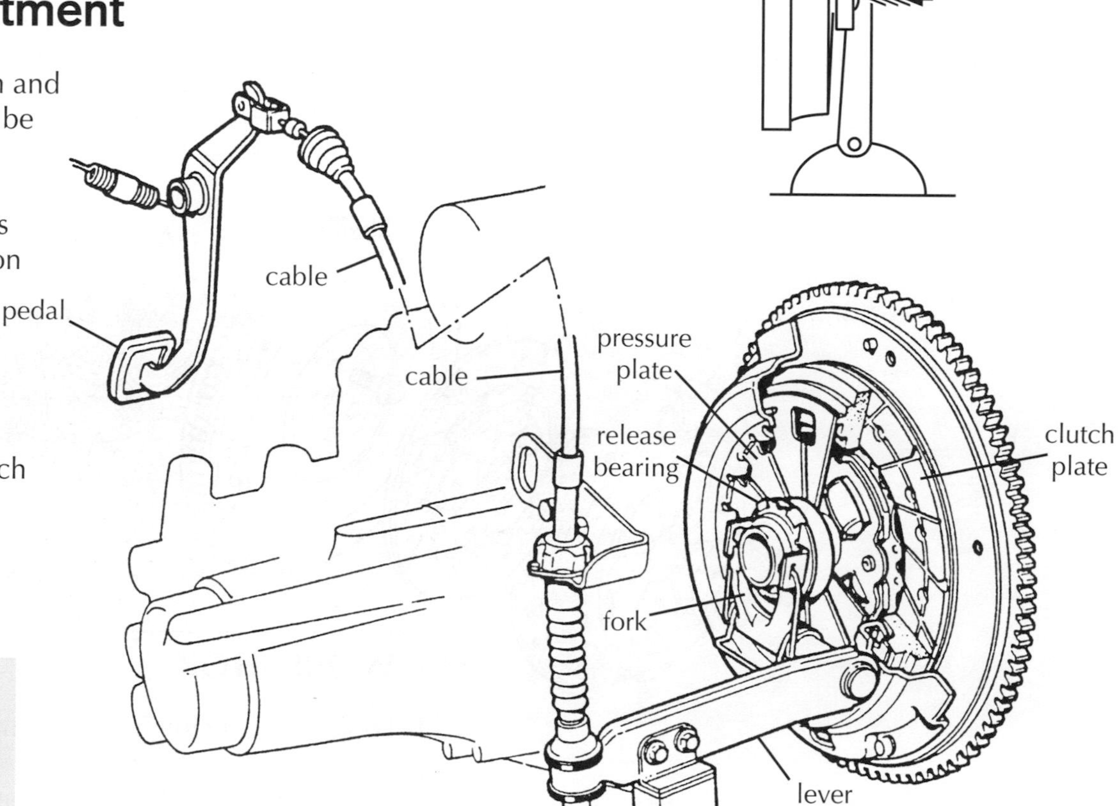

QUESTION

Add an arrow to show the manual adjuster in the cable system.

WORKSHOP

Examine different cars with cable and hydraulic clutch controls. Ask someone to operate the clutch pedal. Watch how the withdrawal lever is moved.

Clutch service and repair

Look at the table below. The first column describes what the driver notices when the clutch is faulty. The second column lists the most likely reasons.

As a rule, clutches only develop faults at high mileage. If there *is* a fault, even if only one part is badly worn, it is usual to renew *all* the major parts of the clutch.

Symptoms	Probable causes	Description
Partial or total loss of drive Vehicle speed lower than normal, compared with the engine speed	Worn clutch linings Oil-soaked linings Incorrect adjustment	**a**
Difficulty in engaging gears, particularly first and reverse gears	Air in hydraulic control system Incorrect adjustment	**b**
Difficulty in controlling the initial drive on releasing the clutch pedal	Broken or loose engine mounting Oil on clutch linings	**c**
Shudder and vibration as the vehicle moves off from rest	Oil on clutch linings	**d**

QUESTIONS

1 Complete the third column of the table by adding the correct description from this list: *drag*; *judder*; *slip*; *fierceness*.

2 On first contact with the clutch pedal, the driver hears a squeal or grating sound. What is the most likely cause?

....

TIP

When removing and replacing the clutch:

- Support the gearbox properly. Do not let it 'hang' on the clutch plate.
- Check the spigot bearing.
- Do not allow oil or grease or any fluid to come into contact with the clutch plate linings.

QUESTIONS

1 Name the parts usually replaced during a clutch overhaul.

....

....

....

2 The tool shown (right) is a **clutch alignment tool**. What is it used for?

....

....

3 The tool shown (below) can be bolted to the rear of the engine during clutch overhaul. Why is it used?

....

....

ACTIVITY

Get information from catalogues or car manuals, or by examining vehicles, and complete this table.

Vehicle		Type of clutch control (cable/hydraulic)	Type of adjuster and position
Make	Model		

Lubrication

The need for lubrication

It is vital to use **oil**. Oil between moving parts reduces friction and wear, and cools the metal surfaces.

In engines, oil also carries away tiny metal and carbon particles. These would otherwise damage the bearings and cylinder surfaces.

A very thin film of oil is all that is needed to keep the metal parts apart

Types of oil

Oils used in vehicles are mineral oils. They are graded by their viscosity and their quality. **Viscosity** is a measure of how easily the oil flows – how 'thick' it is. The *thinner* the oil, the *lower* the viscosity number.

Engine oils

All oils become 'thinner' as they are heated. **Multigrade oils** contain additives so that they 'thin' more slowly. Because of this, a thin oil can be used. This allows easy starting when the engine is cold, and prevents the oil from getting too thin when it is hot.

It is important to use the correct oil for the type of engine being serviced. The examples below show one manufacturer's selection of engine oils, and the reasons for using each one.

ACTIVITY

1 Complete the following sentences:

a Engine oils have viscosity numbers between ....

and ....

b Gear oils have viscosity numbers between ....

and ....

c Automatic transmission fluid is rather different because

....

2 A multigrade oil has two viscosity numbers, for example
5W/40   10W/40   15W/40   15W/50

a Which of these is the thinnest oil? ....

b Which of these is the thickest oil? ....

15W/50 MOTOR OIL
A VERSATILE VALUE FOR MONEY LUBRICANT

- PETROL AND DIESEL
- MINIMUM OIL CONSUMPTION
- MAINTAINS GOOD OIL PRESSURE
- ANTI-SLUDGE QUALITY
- SUITABLE FOR HIGH MILEAGE ENGINES

10W/40 MOTOR OIL
SEMI-SYNTHETIC
COMPLETE PROTECTION FOR PERFORMANCE ENGINES

- SUITABLE FOR TURBO-CHARGED, FUEL INJECTION AND MULTI-VALVE ENGINES
- RAPID PROTECTION FROM COLD START
- ANTI-SLUDGE QUALITIES
- GREATER FUEL EFFICIENCY
- REDUCED ENGINE WEAR

5W/40 MOTOR OIL
FULLY SYNTHETIC
HIGH PERFORMANCE OIL FOR HIGH PERFORMANCE ENGINES

- OPTIMUM ENGINE PROTECTION
- EXTRA POWER RELEASE
- REDUCED FRICTION AND WEAR
- HIGH TEMPERATURE PROTECTION
- STRESS RESISTANT

15W/40 MOTOR OIL
NEW GENERATION ADVANCED PROTECTION FOR FAMILY CARS

- IDEAL FOR NEW FAMILY CARS
- EXCELLENT ENGINE CLEANLINESS
- REDUCES ENGINE WEAR
- MINIMUM OIL CONSUMPTION
- SUPERIOR QUALITY MULTI-GRADE

15W/40 MOTOR OIL
FOR ALL DIESEL CARS AND LIGHT COMMERCIAL VEHICLES

- SUITABLE FOR TURBO AND NATURALLY ASPIRATED
- ENHANCED DIESEL ENGINE PERFORMANCE
- EXCELLENT CLEANLINESS
- REDUCES DEPOSIT FORMATION
- SOOT CONTROL TECHNOLOGY

Lubricants and fluids

Many systems in a car need **lubricants** or **fluids**. A common job for you will be to look under the bonnet and check the **levels** of these lubricants and fluids. Examples are shown in the list below.

On modern vehicles, items that need regular checks often have colour-coded tops.

WORKSHOP

Look under the bonnet of a vehicle. Check whether any fluids are needed.

Fluid	Level		
	High	Low	OK
Engine oil			
Brake fluid			
Clutch fluid			
Radiator coolant			
Battery electrolyte			
Windscreen washer fluid			
Automatic transmission fluid			
Power steering fluid			

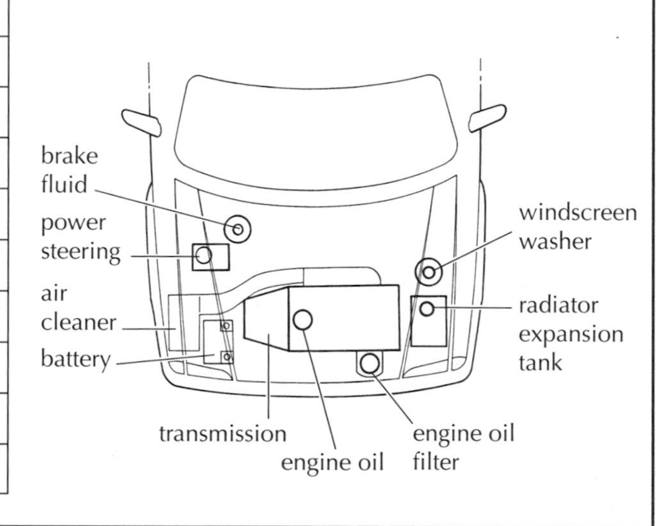

The fluids that are used are very specialised. When refilling, you should always be sure that you use the *correct type* and the *correct amount*.

Check the workshop manual. If you are in any doubt, ask your supervisor.

HEALTH & SAFETY

Always make certain that you are using the correct fluid.

If, for example, you put screen washer fluid in the brake reservoir, the brakes would not work and someone might be killed as a result.

ACTIVITY

Look at this table of engine and component/systems. Using a workshop manual if necessary, state typical lubricants and their grades.

Component or system	Lubricant type/specification
Engine	**1**
Manual gearbox	**2**
Automatic transmission and power steering	**3**
Cooling system	**4**
Braking system	**5**

Fluid capacity

Each system on each vehicle needs a specific amount of the right fluid. This is the **refill capacity**. You will add fluid to reach this *total*. How do you know how much is needed?

ACTIVITY

Look at wall charts or workshop manuals. For a particular vehicle, find out the refill capacity of the different fluid systems.

Vehicle: Make	**Model** ....
System	**Capacity** (litres)
Engine oil	
Manual gearbox	
Automatic transmission	
Power steering	
Cooling system	
Battery	
Brakes	
Windscreen washer	

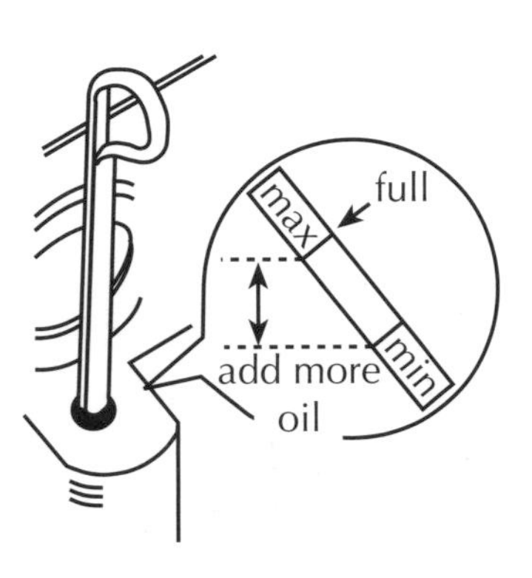

Checking the engine oil level

Engine oil and fuel system maintenance

To keep an engine in good running order, the **engine oil** and the **oil filter** should be changed from time to time, and also the **air cleaner filter** and the **fuel filter** often should be changed at the same time.

Vehicle manufacturers say how often these items should be changed – for example, every 12 months or 12 000 miles. These figures are the **recommended intervals**, but are not the same for every model.

QUESTIONS

1 Name the parts of the engine.

a ....

b ....

c ....

d ....

2 *Two* of these items should be changed at regular intervals. Which two?

....

....

Changing the engine oil and the filter

Oil filters are usually replaceable canisters. They must be replaced when the engine oil is changed. The change period is often every 6000 miles. On cars with catalytic converters and cars that use **synthetic oil**, the change period may be every 9000 or 12 000 miles.

WORKSHOP

Change the engine oil and the filter.

It is easier to do this when the vehicle is raised on a ramp or lift.

Before you start, make sure you have the correct oil, the oil filter, and a new sump plug sealing washer. You will also need an oil filter removal tool, a sump plug spanner, an oil drain tray, and wing covers.

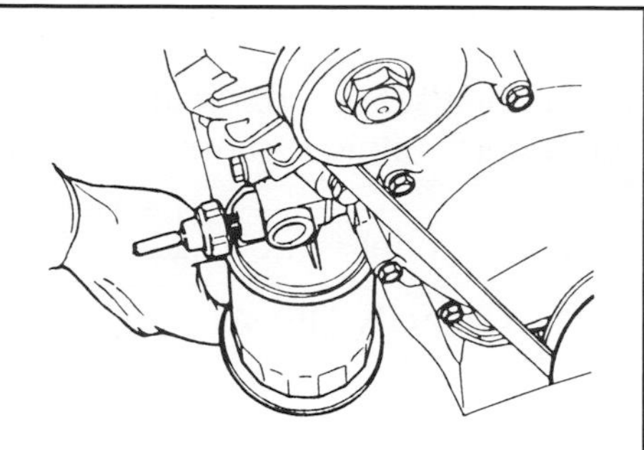

Procedure

1. Position the drain tray.
2. Remove the filter.
3. Drain the engine oil; replace and tighten sump plug.
4. Smear the rubber sealing ring of the oil filter with new engine oil.
5. Replace the oil filter.
6. Fill up with new engine oil, to the correct level.
7. Start the engine and check for leaks.
8. Check the oil level. Top up, if necessary.
9. Clean the vehicle of any spilt oil.
10. Dispose of waste material properly.

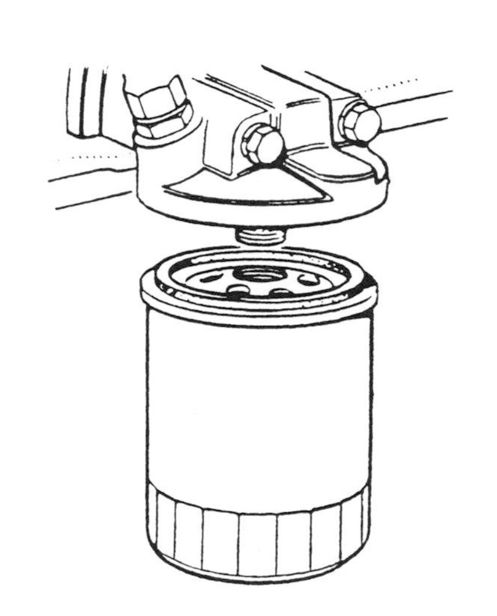

TIP
Changing the oil is easiest when the engine *is warm*. If it is *cold*, the oil will not flow easily. If it is very *hot*, the oil could scald you.

HEALTH & SAFETY
By law, you must not pour old oil into drains, and you must not place oil filters in ordinary waste bins.

Other lubricants

There are three types of lubricants for the **transmission**:

- **Gearbox oil** Some manufacturers use **engine oil**; others use a thin 'extreme-pressure' **gear oil**.
- **Automatic transmission fluid** This is a thin oil needed for the transmission's torque converter, multi-plate clutches and pistons. This type of oil is also used in power steering systems.
- **Final drive oil** In rear-drive vehicles the oil must be an extreme-pressure **hypoid oil**. This sticks to the gears as they mesh together or separate.

Grease is an oil-based substance, a bit like soap. The most common type is a high melting-point **lithium base** grease. This is used for wheel bearings and suspension ball joints.

QUESTIONS

What other types of lubricants are there? Where would you use them?

1 Wax-based ....

2 Copper-based ....

3 Zinc-based..........

4 Silicon-based ....

Engine air cleaner or filter

The **air filter** prevents any dust or dirt in the air from entering the engine. The unit is usually changed at each major service.

The most common sort of air filter is a replaceable porous paper element. This fits into a box at the start of the engine's **air intake**.

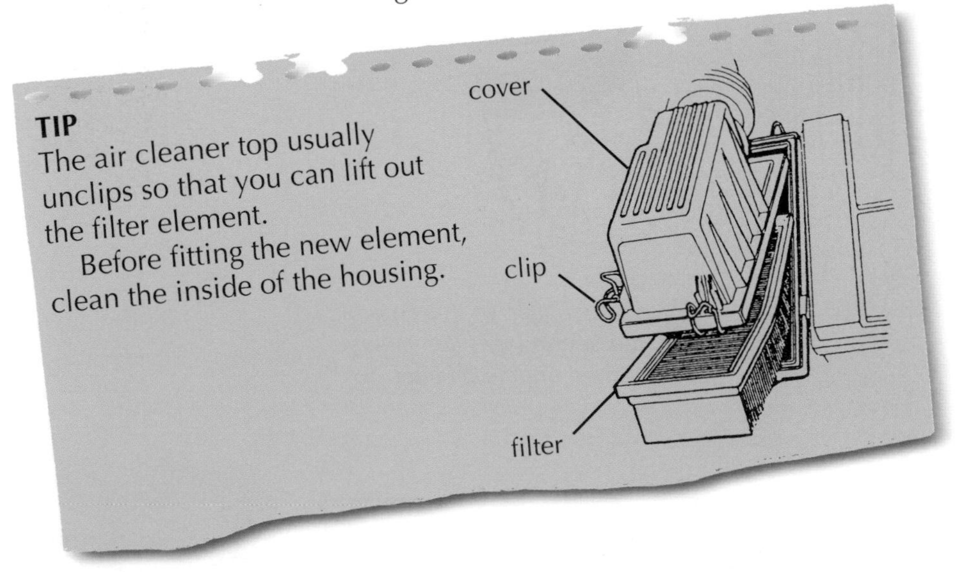

WORKSHOP

Change an air cleaner element.

Fuel filters

The **fuel filter** is usually placed in the fuel line between the tank and engine. It is a pleated paper element.

- If the fuel system uses **petrol injection**, the filter will be a large unit positioned near to the tank.
- If a **carburettor** is used, the filter will be smaller and mounted on the body, near the engine.
- The fuel system for a **diesel engine** is likely to have at least *two* filters: a water trap, and a large paper element.

A petrol injection filter

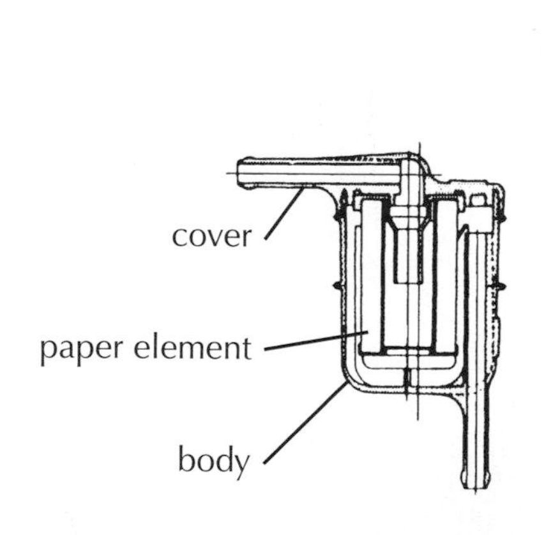

A carburettor fuel line filter

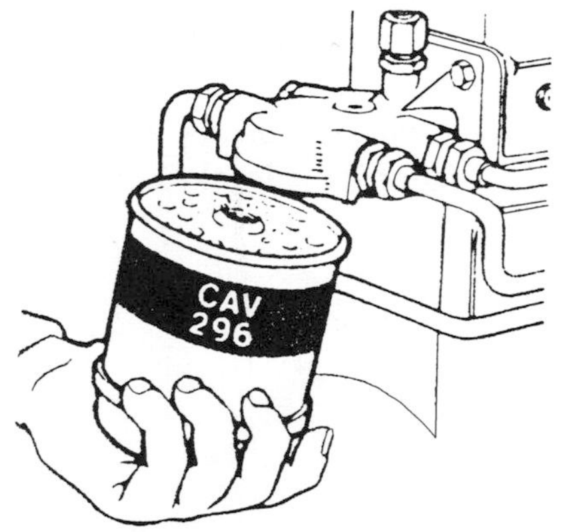

A diesel fuel filter. In this type of filter, the secondary filter is very similar to an engine oil filter

WORKSHOP

Change a fuel filter element.

TIP
When replacing fuel filters, take care that the inlet and outlet pipes are fitted the right way round.

ACTIVITY

Find out the filter change interval for a vehicle in your workshop.

Vehicle: Make	**Model** ....
Filter	**Change interval** (miles)
Oil filter	
Air filter	
Fuel filter	

Modern engines

A 4-cylinder, 16-valve, twin overhead camshaft petrol engine is shown.

Study the engine below, can you identify the main working parts?

Note the items which will/may require servicing such as:

- Oil filter
- Coolant thermostat
- Fuel injectors
- Ignition spark plugs

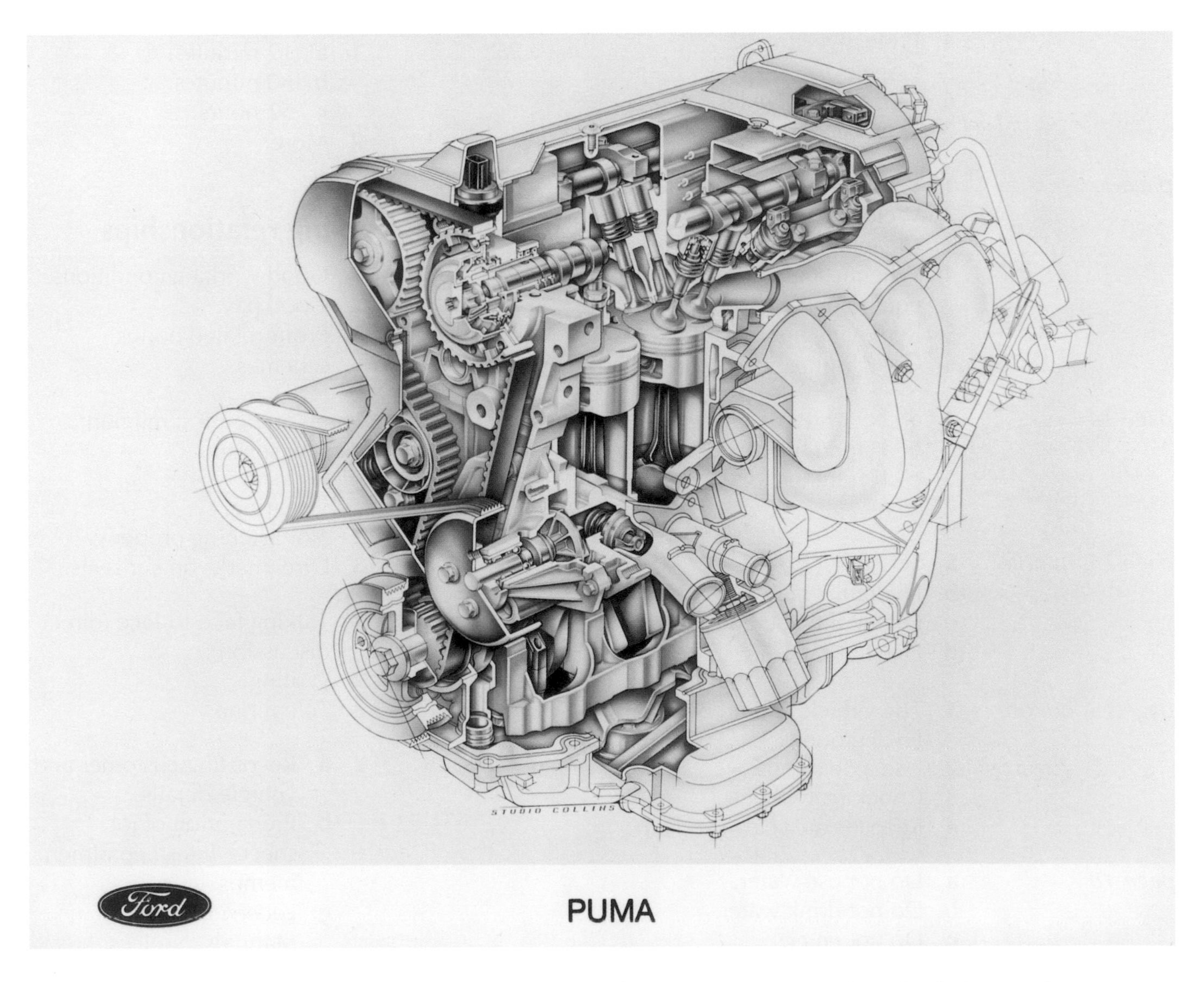

Answers

Chapter 1 Safety and good housekeeping

page 9, middle

1 Wrong clothes.
2 Effects of drink or drugs.
3 An untidy workshop.

page 9, bottom

1 The well-being of the person.

page 11

1 One-piece neatly fitting overall.
2 Tidy, short hair.
3 All buttons fastened.
4 Protective gloves.
5 Goggles.
6 Close-fitting ankles.
7 Steel-capped boots.

page 12

1 Working under a vehicle.
2 Welding.
3 Cleaning dust from brakes.
4 When using noisy body equipment (e.g. sanding).
5 Removing hot exhausts.
6 Car washing.

page 14

1 **a** Fork-lift truck.
  b Hand/sack truck.
  c Flat trailer.
2 **c.**

page 15, middle

a Scissor jack.
b Hydraulic jack.
c Engine lifting crane.
d Hoist or ramp.

page 15, bottom

1 The vehicle should be on level ground.
2 Use axle stands.
3 Chock rear wheels.
4 Keep the area clean and tidy.

page 18

a Do not use water.
b Do not drink water.
c Do not enter.
d Must wear goggles.
e Must wear hard hat.
f Must wear ear muffs.
g Danger: fork-lift truck in use.
h Danger: slippery surfaces.
i Danger: corrosive chemicals.
j Safe way: emergency exit downstairs.
k Eye wash available.
l Fire escape ladder.
m Water hose.
n Emergency telephone.

page 25

1 **a** 12 minutes.
  b 60 minutes.
  c 52 hours.
2 More.

Chapter 2 Working relationships

page 28

1 Good working conditions.
2 Good pay.
3 Profit-related bonus schemes.

page 29, middle

Three of: politics; religion; racism; sexism.

page 29, bottom

1 Swearing.
2 Not listening properly.
3 Dirty marks on car seats.

page 31

1 Talking face-to-face (direct discussion).
2 Writing.
3 Telephone.

page 32

2 **a** Recording customer and vehicle details.
  b Preparation of job sheets. Inter-department memos.
  c Letters. Reports. Manuals. Notices.

3 **a** Quick communication with customers, suppliers and others.
b Between departments, with colleagues.
c Quick contact with emergency services.

Chapter 3 Communication

page 37
1 **a** No name or phone number.
b No name or job details.
c No name or phone number.
2 Loss of communication could mean loss of business.

page 38
1 The firm's security arrangements.
2 Employees' wages or salaries.

page 39
1 112.
2 0800 800 154.
3 192.
4 1471.
5 5.
6 01479.
7 0033.

page 41, middle
1 Looking directly at you.
2 Varies to match what you are saying.
3 Nodding or shaking as appropriate.

page 41, bottom
2 re – regarding.
cc – carbon copy to.

page 43
1 **a** Do not dirty car inside or out.
b Do not place tools on bodywork.
c Make sure job is done properly.
2 **a** Correctly identify faults.
b Technicians do a good job.
c Reasonable prices.
d Clean and tidy workshop.

Chapter 4 Handling and merchandising stock

page 46, bottom
1 This way up.
2 Fragile.
3 Stack 3 high.

page 47, top
1 c.
2 **a** Overhead load.
b Toxic hazard.
c Corrosive substance.

page 47, bottom
Three of: scissors, crowbar, wire cutters, screwdrivers, pliers, hammer.

page 48, top
1 Mounted on a pallet.
2 In moulded polystyrene.
3 In a corrugated cardboard box.

page 48, middle
1 A stick-on label.
2 A painted number.
3 A tie-on label.
4 A sticky label.

page 50
1 First in, first out.
2 Microfiche.
3 It indicates the type of hazard presented by the goods.
4 On lifting equipment. 'Safe Working Load'. (This must not be exceeded when using the equipment.)
5 It is poisonous.

page 51
Anti-freeze, batteries, touring kits, towing equipment.

Chapter 5 Tyres

page 56, middle
1 Speed.
2 Load.
3 Operating conditions.

page 57, middle left

page 57, middle right
1 Cross-ply or diagonal-ply.
2 Radial-ply.

page 58, top
1 Cross-ply.
2 Radial-ply.
Radial ply.

page 58, middle

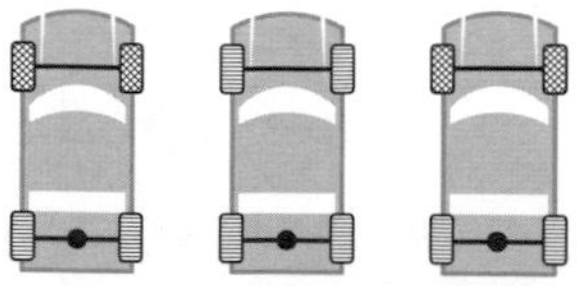

page 58, bottom
1 Looks different.
2 Appears to be thinner.
3 Says 'TEMPORARY USE ONLY' on sidewall.

page 59, middle
185/60 R14 82H.

page 60
130 mph.

page 61, bottom left
475 kg.

page 61 bottom right
Section width (millimetres).
Aspect ratio.
Radial-ply.
Rim diameter (inches).
Load index.
Speed rating.

page 62
Tubed.

page 63
1 Cold.
2 Higher.
3 Check when tyres are cold.
4 **a** Vehicle handbook.
  b Tyre manufacturer's wall chart.
5 **a** **i** Sloppy handling.
    ii Heavy steering.
  b **i** Hard ride.
    ii Steering feels light.

page 65
Non-directional: on right.
Directional: on left.

page 66
Under-inflation: on left.
Over-inflation: on right.

page 69, bottom

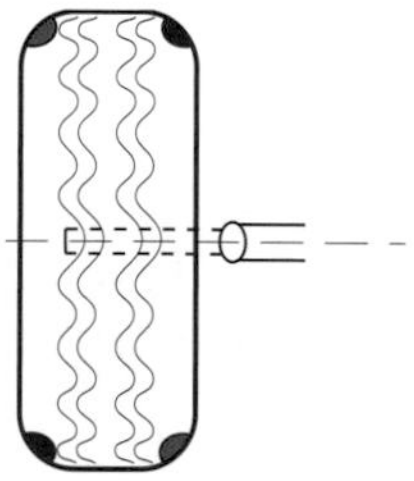

page 70
All answers are correct except the last one.

page 72, middle
1 A nail (or similar object) cutting or chafing the sidewall.
2 The tyre has been run flat, damaging the sidewall.

page 73, top
Every time, without fail.

page 73, bottom
1 **a** Pressed-steel.
  b Alloy.
  c Wire-spoked.

page 74, middle
Inches.

page 74, bottom
Millimetres.

page 75, middle
Rim width (inches).
Flange height code letter.
Rim diameter (inches).

page 75, bottom
1 Bent flange.
2 Deep rust.

page 77, top
1 Detachable flange.
2 Locking ring.
3 Rim base.

page 77, middle
1 Steering could be affected.
2 Twin tyres could touch each other.
3 A tyre could foul the mudguard.

page 79
1 Scooter.
2 Military vehicle.
3 Agricultural vehicle.

page 80

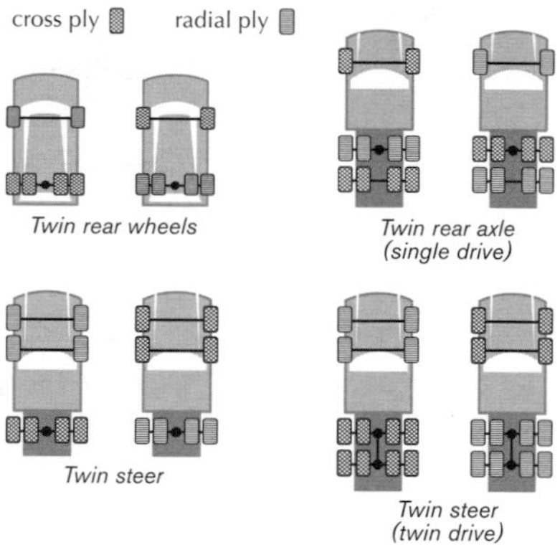

page 81, top
Load: taken largely by only one tyre.
Wear rate: much more on smaller tyre.
Chance of blow-out: much greater.

page 81, bottom left
No.

page 83, top
1 Rust.
2 Dirt or paint.
3 Damage.

page 83, middle
1 Torque wrench.
2 36 lb/ft.

page 84
They are less expensive.

page 85

1 Non-directional.
2 Directional.
3 Non-directional.

page 86

40 kph.

Chapter 6 Electrical

page 90, middle

1 6 cells.
2 3 cells.

page 90, bottom

1 a Negative terminal.
  b Positive terminal.
  c To allow electrolyte to be topped up.
  d Electrolyte level indicators.

page 91

3–5 amps.

page 93

1 b.
2 b.
3 c.
4 d.

Chapter 7 Brakes, suspension, steering

page 100

To apply and release the handbrake when parking.

page 101

To prevent leakage.

page 102

1

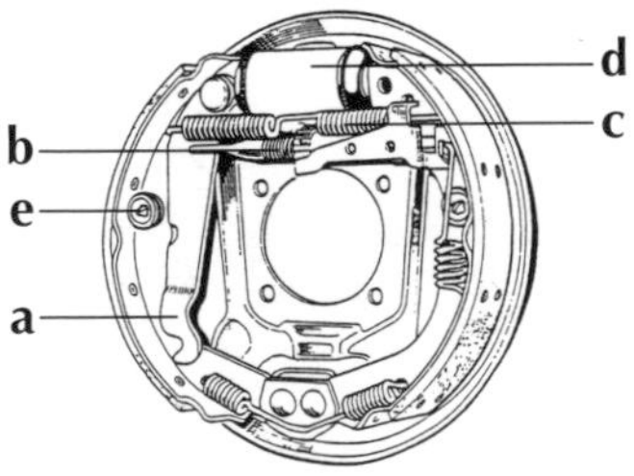

2

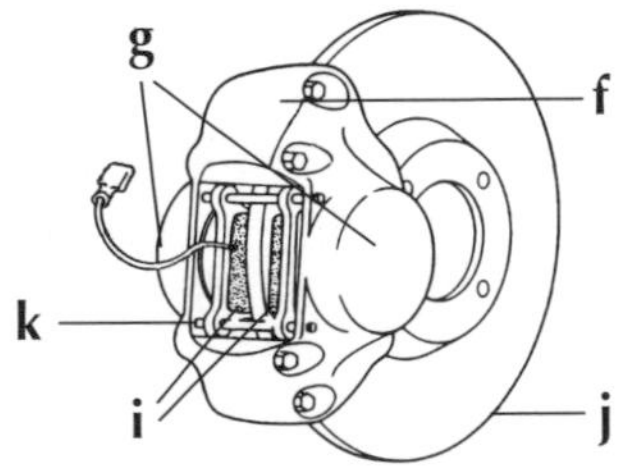

3 To operate 'pad worn' indicator.

page 103, top

1 Excess travel.
2 Excess travel.
3 Efficiency.
4 Fluid level.
5 Pad/lining thickness.
6 Drum/disc scoring or corrosion.
7 Pipe corrosion.
8 Hoses for cracks or other damage.
9 Fluid leakage from wheel cylinders.
10 Frayed handbrake cable.

page 103, bottom

During repair.

page 104

1 Return spring.
2 Pawl and ratchet.
3 Roller brake tester (dynamometer).
4 Footbrake: 50%.
  Handbrake: 25%.

page 105

1 The other system would still operate on two wheels.
2 Servo.
3 ABS (Antilock Braking System).

page 107

2 a Acceleration.
  b Deceleration (braking).

page 108

A Multi-leaf.
B MacPherson strut.
C Fluid/gas.
D Torsion bar.

page 109, top

1 Telescopic.
2 Strut.

page 110, top

1 Fluid leakage.
2 Insecure mountings.
3 Corrosion.
4 Damage.

page 111

1 C.
2 Spring being compressed for safe removal.
3 Anti-roll bar.

page 112, top

1 Kingpins and bushes.
2 Ball joints.
3 MacPherson strut.

page 112, bottom

1 Pinion.
2 Rack.

page 113, top

1 Pinion housing.
2 Steering column.
3 Rack housing.
4 Rubber boot.
5 Track rod end.

page 113, bottom

1 It acts as an oil seal.

2
1. Ball joint.
2. Track rod end.
3. Rubber boot.
4. Boot clip.
5. Boot clip.
6. Nut.
7. Split pin.
8. Lock nut.
9. Track rod.
10. Rack.

page 114 Four.

page 115, top The dust shield.

page 115, middle Ball joint splitter.

page 116
1 Toe-in.
2 Toe-out.

Chapter 8 Exhaust, cooling, clutch, lubrication

page 120
1 Heat shield.
2 Mounting brackets.
3 Expansion box.
4 Absorption box.
5 Gasket.
6 Flexible pipe.
7 Manifold mounting flange.
8 Catalytic converter.

page 122
1 Pipe clamps.
2 Pipe support mountings.
3 Box support – rubber mountings.
4 Box support – rubber block.

page 126
1 b.
2 b.
3 b.
4 b.

page 127, top Through the splines in the clutch plate hub and primary shaft.

page 127, middle By slowly and gradually releasing the clutch pedal.

page 127, bottom
1 C.
2 B.
3 A.
4 D.

page 128

page 129
1 **a** Slip.
b Drag.
c Fierceness.
d Judder.
2 The release bearing.

page 130
1 Clutch plate, pressure plate and release bearing.
2 To centre the clutch plate.
3 To prevent the flywheel from turning.

page 131
1 **a** 5W and 50.
b 70 and 140.
c It is thin and does not have a number.
2 **a** 5W/40.
b 15W/50.

page 133, bottom
1 **a** Air filter.
b Fuel pump.
c Oil filter.
d Sump plug.
2 Air filter and oil filter.

page 135
1 To damp-proof the ignition system; as a corrosion release agent.
2 For screw threads; on brake parts to prevent seizing.
3 For moving metal parts on brakes.
4 When fitting rubber components.

Index